MINDFUL SCEPTICS
GUIDE TO...

biodiversity loss

Mark Dangerfield
Christopher M Scott

Mindful Sceptics

Published by Mindful Sceptics

Mindful Sceptics

Springwood, NSW 2777, Australia

mindfulsceptics.info

AI Use Statement

This book was conceived, written and edited by a human who occasionally used AI tools to assist with grammar, editing, summaries and focused searches.

Contents

Preface to the Series

Ever feel like you're drowning in information and controversy, unsure what to believe? You're not alone. In this age of information overload, finding clarity can seem like searching for a needle in a digital haystack. But what if I told you that the key to navigating this chaos lies within you?

Welcome to the Mindful Sceptic Guides, your compass in the turbulent waters of modern discourse. These aren't your average self-help books or dry academic tomes. They're intellectual adventures designed to awaken your innate curiosity and sharpen your critical thinking skills.

In the pages of *The Mindful Sceptic Guides,* the content compliment to *Being a Mindful Sceptic,* you'll find a delightfully eclectic mix of crucial facts, thought-provoking ideas, and paradigm-shifting perspectives. We'll tackle big issues with a unique blend of scientific rigour and philosophical inquiry. No topic is off-limits, from conservation to social justice, food security to a healthy diet, and economic theory to personal growth.

But here's the kicker. This isn't just about personal enlightenment. The stakes are higher than that. Humanity's survival may depend on cultivating mindful sceptics—people who can question dogma, seek evidence, and imagine new solutions to age-old problems in a world grappling with complex, existential challenges. Could you be one of these crucial thinkers? Could your ideas be the spark that ignites positive change?

There's only one way to find out. Dive into these guides with an open mind and a willingness to challenge your assumptions. Prepare for moments of revelation and disagreements, ideas that will keep you up at night, and insights that will transform how you see the world and your place in it.

Are you ready to embark on this intellectual journey? To become a beacon of reason in a world that desperately needs it? Then, turn the page and welcome to the Mindful Sceptic Guides. Let's change the world, one critical thought at a time.

Preface to this Guide

In September 1996, I arrived at Macquarie University in Sydney, Australia to begin a lectureship role in the newly minted Key Centre for Biodiversity and Bioresources. My job was to teach an undergraduate course in biodiversity and build a research program on the biodiversity of invertebrates. What an honour and a privilege at a time when biodiversity, short for biological diversity, was a new term adopted at the UN Rio Convention in 1992 to refer to the variety of life on Earth, everything from genes to ecosystems. Ambitiously, biodiversity encompasses the diversity within species, between species, and of ecosystems. We are talking about everything from genetics, individuals, populations, species, assemblages of species and even ecosystems; what they all are and what they do. As Professor Ed Wilson put it, the diversity of life. I was super keen to get started. But perhaps I should have thought about what I was getting into.

A single word was supposed to capture billions of years of evolution, an uncounted variety of life forms and what they all did individually and together. Biodiversity was the stuff of life, of nature, and the consequences of that stuff, especially for what came to be called ecosystem services, what biodiversity does for food production, climate regulation, and nutrient cycling. Only, this was just the beginning.

This graphic was redrawn from Professor Andy Beattie's book 'Australia's Biodiversity: Living Wealth' where the size of each organism represents the number of described species in that group.

Because biodiversity did things that meant something to human society and economics, the term was quickly appropriated to cover threats to its existence—habitat destruction, overexploitation, pollution, climate change, and the introduction of invasive species—and the consequences of any disruption to biodiversity for the functioning of ecosystems and the well-being of humans who depend on them. In a flash, biodiversity also became about conservation, especially for the health of our planet and ensuring the continued provision of the benefits we derive from nature. If its already copious scope wasn't enough, it became a synonym for saving species, protecting habitats, managing natural resources sustainably, and reducing the negative impacts of human activities on the environment.

I should have felt overwhelmed by cramming all or even some of this into a 12-week course for undergraduates. I didn't, partly because I simplified the definition to 'multi-species ecology and evolution', making biodiversity a characteristic of nature rather than a product ripe for use, exploitation, and potential loss.

That is not how the world saw it. After a short time biodiversity became conservation, especially species conservation, and biodiversity loss was terrible. It is a problem for humans, perhaps not in the way you think.

This book is about much more than biodiversity loss.

Chapter One

Don't look up

On a finite planet with 8 billion people exploiting the fossil fuel energy pulse, no landscape remains pristine, and pollution is everywhere. Why is this picture so hard to see?

Scientific agreement

I watched the movie *Don't Look Up* and survived. Despite the occasional flat joke and intentionally long runtime, a more visceral take on modern life is hard to find. In the movie, journeymen scientists from Hicksville discover a large meteorite they saw through a telescope and then calculate that it will collide with Earth in weeks. They are ignored, so are their colleagues from the Ivy League colleges who agree and support their findings—not every viewer would appreciate how rare it is to see such scientific agreement. All the non-scientists, from the President down, have way more important things in life than imminent death, except the billionaire telecoms entrepreneur with a space subsidiary who decides there is money to be made off a meteorite full of rare earth metals.

Good satire should be brutal to watch because it is very nearly accurate. If only real life were easier. Here is something else that has similar scientific agreement and is ignored by everyone else.

Elizabeth Kolbert, American journalist, author, and visiting fellow at Williams College, Massachusetts, published a book in 2014 entitled *The Sixth Extinction: An Unnatural History* that won the author the Pulitzer Prize for General Non-Fiction in 2015. The book explores the concept of mass extinctions in Earth's history and argues that the planet is currently in the first mass extinction event caused by a single species and not some major geological event.

Terry Glavin, Canadian author and journalist, wrote a book in 2006 entitled *The Sixth Extinction: Journeys Among the Lost and Left Behind* that combined travelogue and environmental reporting to explore the impacts of extinction and climate change on various ecosystems around the world.

Richard Leakey FRS, Kenyan paleoanthropologist, conservationist and politician, also wrote a book published in 1995 entitled *The Sixth Extinction: Patterns of Life and the Future of Humankind* that explored past mass extinctions and their relevance to understanding the current biodiversity crisis.

Here are a few more books on the topic:

- *The Annihilation of Nature: Human Extinction of Birds and Mammals* by Gerardo Ceballos, Anne H. Ehrlich, and Paul R. Ehrlich

- *The Future of Life* by Edward O. Wilson

- *The Song of the Dodo: Island Biogeography in an Age of Extinctions* by David Quammen

- *The Last Giants: The Rise and Fall of the African Elephant* by Levison Wood

And on it goes.

Then, if you pop across to Google Scholar, that's the academic research tool that searches only the scientific literature, and type 'sixth mass extinction' into the search bar, more than 260,000 records are recalled, with the top few hundred presenting detailed research with the term in the article's title. The first scientific paper on the list is a significant review published in 2022 by researchers from the University of Hawaii, the Sorbonne Université, and the Muséum National d'Histoire Naturelle, Paris, that cites 245 sources and makes statements like this one.

> *As an example, we focus on molluscs, the second-largest phylum in numbers of known species, and, extrapolating boldly, estimate that, since around AD 1500, possibly as many as 7.5–13% (150,000–260,000) of all ~2 million known species have already gone extinct, orders of magnitude greater than the 882 (0.04%) on the Red List.*
>
> Cowie, R. H., Bouchet, P., & Fontaine, B. (2022). The Sixth Mass Extinction: fact, fiction or speculation?. Biological Reviews, 97(2), 640-663.

Just pause on this for a moment. That is at least 150,000 species extinct at a minimum. Compare that number with all the fuss over saving the koala that you can read about in *The Mindful Sceptic Guide to Saving the Koala*, and then consider this advocacy from the same scientific paper on mollusc extinctions.

> *We consider that the Sixth Mass Extinction has probably started and present arguments to counter those who would deny this. We will probably not convince those who consider this episode part of the natural evolution of life on Earth and therefore that it is acceptable to just let it happen, that is, that they are wrong. Nor will we convince those who think that it is too late to stop it and that we should therefore embrace it, that is, that they also are wrong. However, we hope that we will at least give pause to those who by denying or downplaying it play into the hands of those who advocate doing nothing about it, or those who, accepting it, advocate that we should do our best to manipulate biodiversity primarily if not solely for human, essentially economic, benefit.*

The sceptic would expect the scientists to be cautious. They are poking around in the sand for molluscs and, steeped in their innate objectivity, would be keen to show us the facts. Then, I came across this snippet. According to the largest reinsurance company in the world, Swiss Re, more than half of the world's annual GDP, some $42 trillion, depends on high-functioning biodiversity, and a fifth of the world's countries may see their ecosystems collapse. This sounds like more than an academic search for variety in life; it is a substantial risk.

Failed targets

Our international efforts to deal with biodiversity loss are not much better. The UN has reported that the world has failed to meet a single target agreed upon a decade ago to stem the destruction of wildlife and life-sustaining ecosystems. The UN's overarching plan is for humanity to live in harmony with nature by 2050 and the 2030 goals relate to that ambition in five ways. First, ensure no net losses in the integrity and size of freshwater, marine, and terrestrial ecosystems. This means protecting and restoring habitats to maintain the health and diversity of these ecosystems. Second, reducing the number of species threatened with extinction which involves implementing conservation measures to protect endangered species and their habitats, as well as reducing the threats that contribute to their decline. Third, genetic diversity can be enhanced by maintaining the variety of genes within species, which is essential for their ability to adapt to changing environments and for the resilience of ecosystems. Fourth, achieving the targets of the Paris Agreement. This involves taking action to reduce greenhouse gas emissions and limit the impacts of climate change on biodiversity and ecosystems. Finally, sharing the benefits of genetic resources and traditional knowledge. This means ensuring that the benefits derived from the use of genetic resources and the knowledge of indigenous and local communities are shared fairly and equitably.

Targets like these and much of the political rhetoric associated with them are problematic. They are set up for failure because they do not explain how they can be met. How do you achieve no net loss or enhance genetic diversity? Are these outcomes even possible?

We know that biodiversity—the variety of life on earth—is central to how nature functions because organisms create biomass, move it around and continue to do so even as humans manipulate a huge chunk of it into food, fibre and fuel. As biomass is created by the plants and moved around by the animals that eat them and the microbes that decompose the leftovers, a raft of other 'useful to humans' things happen, such as oxygen production, carbon sequestration, air and water filtering, water harvesting and Uncle Tom Cobley and all. We also know that these critical functions are mediated in large part by soil biodiversity.

Chip away at biodiversity with some loss here and there, and not much changes until it does and systems collapse. All the ecological scientists and the research papers that they write tell us that diversity is crucial to resilience and the ongoing productivity of natural and agricultural systems. They have looked down, seen the biological equivalent of the comet and calculated its trajectory to the sixth mass extinction.

By the way, *Don't Look Up* was about carbon pollution forcing irreversible and catastrophic climate change. It needed a mountain-sized meteor analogy for a satirical edge. God only knows what analogy is needed to get traction for the even more obscure yet equally terrifying risks from biodiversity loss.

Silent Spring

On 27 September 2022, it was 60 years since the publication of Rachel Carson's Silent Spring, an early warning of what human economic activity will do to nature.

> *I don't like the birds singing. It's not my thing.*
>
> A real housewife of Miami

Silent Spring was seminal. Google Scholar has the book cited in 22,490 articles in the last 20 years alone, and a full Google search generates 165,000,000 results. People know this book. Many of them even knew what it was about—something to do with humans making a mess of the environment with pesticides, a problem we fixed by banning DDT.

Yes, organochloride pesticides were banned in most of the world, which was good. But it was nowhere near enough. Even a peripheral knowledge of the evidence tells you that nature conservation has failed.

Since 1962, international conventions have been signed, environmental protection agencies have been established, national parks gazetted, a whole industry of environmental NGOs has emerged, and even a new word has been used to describe it all: biodiversity.

And here is what has happened

- The World Wildlife Fund reports a 60 % decline in the populations of mammals, birds, fish, reptiles, and amphibians in just over 40 years

- Research using conservative assumptions estimates the average rate of vertebrate extinctions over the last century is up to 100 times higher than the (mainly pre-industrial) background rate.

- Insects are declining in both numbers and loss of populations, with long-term monitoring studies showing the number of invertebrates that fly have declined by 76% in 23 years.

- Seabirds have undergone a 70 % community-level population decline between 1950 and 2010 because humans are better at catching so much of their food.

Again, we are heading for a silent spring.

Why has biodiversity declined?

Like a large comet on a collision course, it should be obvious why biodiversity has declined. Humans have appropriated nature. Help has come from a huge army of fossil fuel slaves that gave the power and the machines to do the heavy lifting and allowed humans to use almost all the ecological space required by other species.

Ecologists call this competitive displacement and suggest it is the largest factor in biodiversity loss. *Homo sapiens'* expansion has eliminated 83 % of wild animal and 50 % of natural plant biomass, most of it in the last 100 years. Today, human bodies are 36 % of the mammal biomass on earth, and our domestic livestock another 60 %, leaving just 4 % for all wild mammal species combined. Our poultry accounts for 70 % of Earth's remaining avian biomass. These are numbers that should make you think. If you prefer a visual, try this one.

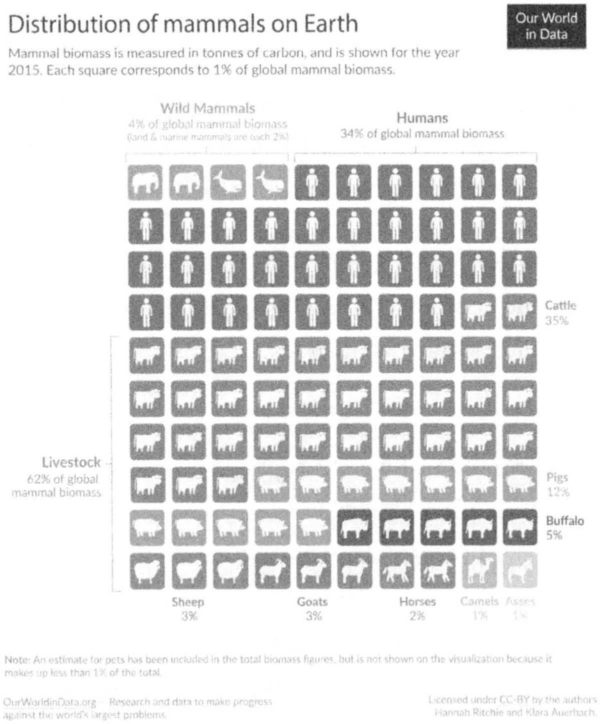

How did we manage to appropriate nature? A combination of fossil energy driving the mechanical slaves, access to the land and resources of the 'new world', and improvement to primary health care suppressed natural negative feedback on our death rate. It freed an innate capacity to expand exponentially both in numbers and space. All organisms do this when presented with an energy surplus and no other constraints on numbers. What happened was that humans got good at competitive displacement and population growth; really good.

Rachel Carson alerted everyone to the potential for mass extinction from pollution. It was a critical contribution. Three generations later, there are 3 billion more people, the consequence of competitive displacement, and the prophecy is coming true as our economic activity exceeds anything imaginable in the 1970s. And for way too many of the 8 billion people alive today, it doesn't even matter. They don't know it is happening and are too busy making ends meet.

Australia uses approximately 6,000 tonnes of active pesticide ingredients annually

The invisible big picture

I've often wondered why the big picture is so hard to see. It is odd because the crisis stories, from climate change to biodiversity loss to growing enough food, are easy enough to explain, and the reasons are obvious enough: 8 billion people, all with the desire to live like kings. It's classic Blind Freddy—the imaginary person supposed to have little or no perception. What Blind Freddy can understand must be very obvious.

Pause on biodiversity loss for a moment. Take a breath, then take in the reality. Here it is.

Biodiversity is in decline because humans have appropriated one-quarter of net primary production (NPP). If humans use one-quarter, there is less biomass available for non-human organisms, and what is there is in a more simplified form, so there is not enough to share around all the organisms on the planet when most of it is channelled for the use of one species. And this appropriation is a trend; human appropriation of NPP is rising steadily.

> *HANPP has risen from 6.9 Gt of carbon per y in 1910 to 14.8 GtC/y in 2005, i.e., from 13% to 25% of the net primary production of potential vegetation. Biomass harvested per capita and year has slightly declined despite growth in consumption because of a decline in reliance on bioenergy and higher conversion efficiencies of primary biomass to products. The rise in efficiency is overwhelmingly due to increased crop yields, albeit frequently associated with substantial ecological costs, such as fossil energy inputs, soil degradation, and biodiversity loss. If humans can main-*

> *tain the past trend lines in efficiency gains, we estimate that HANPP might only grow to 27–29% by 2050, but providing large amounts of bioenergy could increase global HANPP to 44%.*
>
> Krausmann, F., Erb, K. H., Gingrich, S., Haberl, H., Bondeau, A., Gaube, V., ... & Searchinger, T. D. (2013). Global human appropriation of net primary production doubled in the 20th century. Proceedings of the national academy of sciences, 110(25), 10324-10329.

Only it's worse than just the appropriation that might be a third or could reach one-half. As researchers from Rutgers, The State University of New Jersey, established and reported in PNAS, humans and their livestock make up 96% of the mammalian biomass. Appropriate a quarter of the fundamental unit of existence—plants capturing the sun's energy through photosynthesis to produce fuel that animals respire—and loss of diversity is inevitable. If one species takes over resources and channels them into food for itself, eventually, there is not enough for the other species. Oh yes, and the appropriation process uses energy and materials that readily pollute the air, land and water, further stressing the organisms that have just had a bunch of their food source and living space appropriated.

Blind Freddy?

Once humans figured out that ancient energy captured by plants was stored in the deep layers of the earth's sedimentary rocks and that drills or mine shafts could bring that energy out of the ground, well, they went for it. Meanwhile, it was obvious to serious scientists that the composition of the atmosphere would change with all that burning of ancient plant material. And when the products of those burns are the gasses that keep the planet warm, then surprise, more energy is retained in the atmosphere and the climate changes. A staggering 'no brainer'. Again it is as though we can't accept the obvious. Perceptionless Frederick again.

Each day the global economic engine of humanity uses 97 million barrels of oil.

While we have Freddie in the room, we might ask him what he thinks would happen if one species gathers a huge chunk of the available food and is supported in that effort by fossil fuel slaves and kick-ass technology.

More people, dumbass.

Please leave it to the youngsters

It's remarkable how little most of us understand about the world around us. People are no longer curious and have lost their innate abilities to understand and connect with nature. As we've become cocooned in our homes, glued to our drama augmented by binged episodes of Real Housewives, the requirement to understand where all of it came from and whether or not we need to do anything to keep it going beyond getting up and going to work seems to have passed a lot of us by.

And the old dogs who have spent a lifetime enjoying the hard work of the fossil fuel slaves and having dominion to learn new tricks. They bark at you.

Approximately 4.4 billion people lived in urban areas in 2021

The encouraging part is that the youth may be paying attention. Many are railing against climate issues and protesting for their future. They want changes so they can have opportunities in the future. Somehow, they've twigged that their world might not be as great as the one their parents inherited and certainly not as great as the one their grandparents looked forward to.

This is an encouraging sign. As they channel that passion into protest and try to get the messages heard by the stupid white men they are about to replace, we hope they channel some energy into curiosity. The solutions to global problems will come from the imagination of the youth. They can come up with ideas that are practical, rational and achievable. And this means they need to get their heads around how the world works. Build the capability to understand so they can filter their own ideas and not depend on the historical thinking of the people who got us into this mess.

But the kids are up against it. The stupid white men are not just going to fade away. For example, it wasn't very reassuring that one of the unheralded items in the Australian federal budget 2020 was a significant reduction in the resources provided to environmental science degrees in Australian universities. The government has promoted STEM subjects and chosen to reduce its input into the environmental sciences. Now, we could unpack this and argue that environmental scientists are probably to blame for this, failing miserably to raise the stakes politically. But when education and awareness are needed to prime the imagination for real solutions, the government sends another poor message to the youth who want to get trained and educated in these environmental solutions.

The mindful sceptic sees all this and does it anyway by firing up curiosity and engaging with critical thinking. The sceptic educates themselves on the scientific method, learns how to pick the evidence from the bullshit and gains some experience in how resource use impacts and interacts with biodiversity.

Individuals paying attention to biodiversity loss have a rewarding and successful career ahead of them because biodiversity loss is a huge risk, accompanied by opportunity.

Key Points

- Scientific evidence strongly suggests that we are in the midst of a sixth mass extinction event caused by human activity, yet the general public and policymakers are largely ignoring or downplaying the severity of the situation.
- The primary reason for the decline in biodiversity is the human appropriation of nature and resources, driven by fossil fuel energy, access to new lands, and improvements in healthcare, leading to a massive increase in human population and livestock at the expense of wild animals and plants.
- Despite early warnings and increased awareness, such as Rachel Carson's "Silent Spring," international efforts to address the destruction of ecosystems have failed to meet targets, and biodiversity continues to decline.
- The younger generation can channel their passion and curiosity into understanding the world and developing practical solutions to address the biodiversity crisis by educating themselves, engaging in critical thinking, and paying attention to the impact of resource use on biodiversity.

In the next chapter

As I watched a snail slowly making its way across a tarmac road, I couldn't help but wonder—should I intervene and save this tiny creature from almost certain death, or should I let nature take its course in the new world we've created?

Chapter Two

Why did the snail cross the road?

Why indeed? A snail has no notion of roads, what they are made of, what they are used for or even that it is a lousy habitat for a gastropod. So what? It's just a snail.

A changed environment

Snails are uncommon in the Blue Mountains west of Sydney, Australia, where I live. The snail I see crossing a tarmac road on my morning walk is unusual. I've never seen one quite like this one before. The sandstone-derived soils of the region lack calcium and are exposed to volatile moisture conditions that don't suit gastropods. This particular snail was small and had just started on what, for it was an epic journey across the tarmac. I'm tempted to pick it up and return the errant creature to the bush for a better chance of survival. The question is, should I move it? Should I save this individual organism from death under a passing car or from dehydration by returning it to its natural habitat? Or should I leave it to its marathon crossing?

This is the conundrum we have faced since humanity altered the planet. Hundreds of thousands of species' habitats are disrupted due to our presence—a change that affects behaviours and survival probability. I'm tempted to save that individual snail. Still, the reality is that the survival probability of the snail species it represents is determined by its adaptation to the new environment that humans have created, not by any random act of a morning walker.

Plus, this new environment created by humanity is not going away. Human activity is a massive disturbance to the natural systems of the world. Species that survive such a disturbance must be able to live with us. Anything that can't has a very high probability of extinction. There are very few wild areas left, and most of those wild areas are essentially non-livable. In other words, most biodiversity struggles to live there, too, because these places are either too cold or too dry or bereft of nutrients.

There are thought to be around 2,500 to 3,000 species of land snails native to Australia. A large percentage of these species are endemic, meaning they're found nowhere else in the world.

At some point, we have to accept this as a reality. Humans appropriating the planet have massively changed it. And the priority for those humans will always be themselves. Feeding ourselves to maintain our health and well-being will always be our priority. We won't sacrifice ourselves for other species. We may tell ourselves that we will, but that's not what will happen when the push comes to shove. Humans will prioritise themselves over other species. Biodiversity is and will be lost. Only there is an irony here. Humanity's huge challenge is that if too many species are lost, it will endanger our own survival.

Ecosystem services

The goods and services that humans need from nature are the consequence of other species. Some we eat. Some we use to build our homes or fuel our fires. There are those that we admire. Then, there are all the species that must be present to clean the air and the water, move nutrients around, and make life possible. These species moderate and deliver critical ecosystem services.

Rather than call it nature, the environment of physical and biological inevitabilities where we once foraged, hunted, sheltered, lived and died, we now call ecosystem services. Nature in service of humans through the direct and indirect benefits that humans obtain from ecosystems and the natural environment. We identify these services as crucial for the well-being and the functioning of our societies and economies. Naming them as services, we assume, will highlight their value and the importance of protecting and sustainably managing them.

Academics have passed on to policymakers four categories of service. *Provisioning Services* are the products obtained directly from ecosystems, such as food, fresh water, timber, fibre, and medicinal resources. *Regulating Services* are the benefits obtained from the regulation of ecosystem processes, such as climate regulation, air and water purification, pollination, pest and disease control, and natural hazard regulation. *Cul-*

tural Services are the non-material benefits people obtain from ecosystems, such as recreational opportunities, aesthetic experiences, spiritual enrichment, and cultural heritage. The fourth category of *Supporting Services* is a catchall for the processes that allow for all other ecosystem services, such as nutrient cycling, soil formation, and primary production (photosynthesis).

Some say that this formulation makes it easier to understand nature, to measure it and to better understand and communicate nature's value in human well-being. It has been used to inform decision-making in fields such as conservation, natural resource management, and land-use planning. Others argue that it can lead to the commodification of nature and assume an intrinsic value of ecosystems and biodiversity. Only we don't have a mechanism for valuation. For the most part, ecosystem services are not traded in markets and are difficult to quantify in monetary terms. Instead, we trade the products of the services, crops, livestock, timber, even the carbon in the trees.

The reality and the challenge is that biodiversity, the diversity of life, delivers and moderates these services. The common species do most of the heavy lifting and the rarer ones chip in. Only it's not just one or two but thousands of species in complex combinations and webs of interaction that keep the services supplied.

How many species are needed to maintain ecosystem services is a complex and context-dependent question. Generally, more diverse ecosystems tend to be more stable and resilient, providing a wider range of ecosystem services. However, the relationship is non-linear because some species contribute more than others. There are species, sometimes called keystones, that have such disproportionate impacts on ecosystem functions that their loss can lead to significant changes in ecosystem services, even if overall species numbers remain high. Some species may perform similar roles in an ecosystem, a redundancy that can provide a buffer against species loss, but it also means that only some species are equally critical for maintaining services.

Context matters in terms of the number of species needed. For example, a smaller number of key pollinator species might maintain pollination services in an agricultural setting, while soil fertility in a forest ecosystem might depend on a much larger and more diverse community of soil organisms. Research has shown that multiple species, often 10-20, are typically needed to maintain full ecosystem functionality, even in simplified experimental systems. This number is likely much higher in natural systems due to increased complexity and environmental variability. Variability of context matters because higher biodiversity protects against environmental fluctuations, maintaining ecosystem services over time.

While wheat fields are often viewed as monocultures, they still rely on a variety of species to function effectively. While the main crop is wheat, modern agricultural practices often incorporate crop rotation or intercropping, which can involve 2 to 3 crop species over time. A healthy wheat field soil typically contains thousands of microbial species, including bacteria and fungi, with at least 10 to 20 key microbial functional groups that are crucial for maintaining soil health and nutrient cycling. Although wheat is wind-pollinated, surrounding areas often benefit from 5 to 10 pollinator species to support other crops and wild plants. Typically, 5 to 10 species of predatory insects (e.g., ladybugs, lacewings, parasitic wasps) and 2 to 3 insectivorous birds contribute significantly to pest control within the crop. In the soil, earthworms and other soil invertebrates are crucial for soil structure and nutrient cycling with dozens of species even in a teaspoon of soil. A functioning wheat field ecosystem typically relies on at least 50 to 100 species to maintain key ecosystem services like nutrient cycling, pest control, and soil health.

So, how many species can we lose? Written in the research literature, sometimes with great vigour and huge catastrophe, other times more subtly, is the reality that **we don't know exactly how many other species are needed in landscapes for humans to persist**. Along with limited knowledge of how many species we need, there are policies or practices to ensure that a sufficient number of species survive to supply us with the goods and services we require.

The imperative still is to create commodities for sale. The food supply system prioritises production and production efficiency over the services provided by organisms that, as far as the farmer knows, they don't ever use.

Profitability and production efficiency are much more easily understood with inputs of fertilizer, pesticides, and tillage rather than understanding the more subtle and complex delivery of services from nature.

A Paradigm Shift in Food Systems

Humanity needs a fundamental reimagining of our food supply systems. We must elevate ecosystem services to at least an equal footing with profit. This change will inevitably curtail profitability in some areas to ensure the longevity of production through maintained ecosystem functions. This long-term perspective will clash with the short-term profit imperative central to capitalistic models, prioritising the rapid conversion of natural capital into financial returns. Our challenge lies in crafting a new approach to food and fibre production that either diverges from profit-driven motives or finds a way to generate profit without compromising the organisms providing critical long-term services.

First, what do I mean by 'long-term'? At its core, this concept encompasses the persistence of supporting and enabling ecosystem services far beyond the lifespan of any individual farmer or agricultural enterprise. Long means across generations, time enough for the demographic transition to stabilise numbers and demand. It is beyond the business horizons of yield, returns and even passing the business on to the kids.

We know how to do this. It starts with a deep dive into the specific services each ecosystem provides and the mechanisms behind them. Researchers have uncovered how ecosystems function and placed it in the vast repository of scientific literature—our collective knowledge base. Distilling this knowledge into practical solutions is crucial, highlighting gaps in our understanding and guiding future research efforts. It's remarkable how much insight can be gained with the right focus and the ability to distinguish evidence from conjecture.

The next phase involves mapping our food production systems and analyzing farmers' methods for producing commodities for consumption or sale. We must evaluate these approaches not just for immediate efficiency and profitability but also for long-term productivity. How long can a field yield a crop or a paddock rear livestock? True efficiency in food production can't come from depleting soil carbon reserves; instead, it must arise from aligning production with each locality's natural nutrient cycling capabilities. Our current agricultural systems often rely heavily on energy and nutrient subsidies, which we must manage in closer harmony with local ecosystem capacities.

While intensification remains a critical component of global food production, we can no longer assume it will deliver indefinitely without disrupting core ecosystem services. Our path forward lies in understanding how production can work in concert with natural systems, rather than against them or as a substitute.

This shift demands a nuanced comprehension of the intricate interplay between agriculture, ecology, and economics. It calls for new economic models that can fully account for the value of ecosystem services and biodiversity. By embracing this holistic, long-term perspective, we can work towards a food production system that meets our immediate needs and safeguards the ecological foundations upon which it depends.

Why did the snail cross the road?

This chapter began with a snail crossing the road. From there, you can see how easy it is to wander off into the much bigger picture of how and even if humans will persist on the planet.

Whether to save that snail from near-certain death on the busy road by replacing it back into its natural habitat is a critical one. It is fundamental to how we have to operate now that we've altered every landscape on Earth. Knowing how many species we need to survive and if the snail is among them is a determinant of our future.

But we need to understand that the snail crossed the road because it had no choice. Humanity took away its choice by building the road.

Key Points

- Human activity has dramatically altered the planet, forcing species to either adapt to the new environment or face extinction. The snail crossing the road symbolizes this dilemma and the limited choices available to organisms in human-dominated landscapes.

- Ecosystem services, the benefits humans derive from nature, are crucial for human well-being and the functioning of societies and economies. However, the concept raises concerns about the commodification of nature and the difficulty in quantifying the value of these services.

- Humanity must shift food supply systems to balance ecosystem services and profitability to ensure long-term sustainability. This requires understanding the long-term implications of resource use and finding a balance between production and the natural capacity of local ecosystems.
- Addressing the biodiversity crisis requires mining scientific literature to gather information, identify knowledge gaps, and map food production systems to assess their efficiency and sustainability. Understanding the importance of each species in the context of human survival is crucial for determining our future.

In the next chapter

It could be that humanity's fate rests on insects and microbes, the hidden soil heroes. The mindful sceptic must confront the stark reality of biodiversity loss beneath our feet.

Chapter Three

The insect apocalypse and ways to avoid it

Insects and their spineless relatives are too essential to let them suffer an insect apocalypse.

Bugs, yuk!

An insect apocalypse is a good thing, right? No more termites eating through the rafters or spiders sending young girls into screaming fits. No more pesky bugs in your ear, mosquitoes biting your ankles or cockroaches in the kitchen. Butterflies are lovely, but bed bugs from your AirBnB are not. On balance, most city dwellers would vote for an insect apocalypse.

Of course, a moment's thought and these instinctive responses fall over. Have you any idea what fish, lizards, amphibians, and bats eat? Did you know that insectivorous birds consume 500 million tonnes of prey annually? Or that each year, somewhere between $235 and $577 billion

(U.S.) worth of annual global food production relies on pollination from honey bees, native bees, and flies? And this is not half of it. Tiny creatures control pests, recycle organic material from dung to corpses, tree trunks, and leaves, keep the soil healthy, disperse seeds, and contribute to many ecosystem services. All of this happens through the existence and actions of insects and their invertebrate cousins, which we call bugs.

In 2018, Alloporus reported on long-term monitoring studies showing that the number of invertebrates that fly had three-quarters of their abundance gone in a human generation. Here is a more recent quote from an article by Dave Goulson, Professor of Biology at the University of Sussex and the author of *Silent Earth: Averting the Insect Apocalypse.*

> *It should thus be of deep concern to all of us that insects appear to be disappearing from the earth at alarming rates. In Germany, flying insects have declined by 76% in just 26 years. In the UK, our more common butterfly populations have fallen by 46% since 1976, and the rarer ones by 77%, despite great efforts to protect them by conservationists. Thirteen bee species in the UK have gone extinct and more will probably follow. In the US, the celebrated monarch butterfly, famed for its annual migration between Mexico and Canada, has declined by more than 80% since the 1980s. The monarch population west of the Rockies is down 99.9% in the last couple of decades and looks headed for extinction within a year or two.*
>
> Professor Dave Goulson

Insectageddon, as some have dubbed this disaster. Three quarters. One half.

These are huge proportions. Imagine if you lost 75% of your salary or half your superannuation. Panic would ensue, and action demanded with shots fired at whoever might be responsible for such a tragedy. But this loss of bugs barely registers. The newsfeeds make it a filler somewhere down the list, with few people even realising what it means.

Significant loss of insects means that human impact on the environment is now acute. Land clearing, pollution, global warming and the channelling of net primary production into human food have crippled the hardiest creatures, organisms that have been around for 350 million years. And there will be consequences. Insects are not just food for birds and bats; they are integral to the ecological balance, the processes that sustain plant life, and the likelihood of human persistence. I kid you not.

Remember that most of the food required by the global population of humans to supply the 22 trillion kilocalories a day is grown in soil. That only happens because soil biodiversity, the organisms that live in soil, facilitates the exchange of nutrients into plant roots. Lose the soil animals, and the soil becomes inert. Soon after, agriculture collapses.

It is time to register such events with more than a passing 'oh wow' and take action to halt the loss. We must look closely at insecticides and the leakage of fertilisers into waterways and native vegetation around fields. Land-clearing decisions must be more intelligent. Clearing native vegetation shouldn't be because a developer will make a dollar or a landholder says that food can be grown if the trees are cut down. We must know these claims are valid and make informed decisions on the value propositions.

Insects are critical to decomposition and nutrient cycling; they are also a nitrogen and protein source. Instead of eradicating them, cultivating promising species might be an idea.

We might think about restoring habitats not just for cute and cuddly critters but also for invertebrates. The handy part of this is that it is much easier to regenerate insect populations than tigers or Leadbeater's possums.

It is past time landscapes were more than something to paint or use as a selfie backdrop. We might recreate a matrix of vegetation that can support a diversity of insect species in functional patches rather than corridors for critters that never use them.

Shift a big chunk of conservation effort from the rare species to the common ones that do all the heavy lifting for ecosystem services because biodiversity is not a synonym for conservation.

Before embarking on these essential actions, we might raise awareness of how nature works and the critical role of invertebrates. Bugs are yucky, and some are even dangerous to humans. There are poisonous spiders, scorpions that sting, bull ants that sting and bite, and other species that transfer parasites or are disease vectors. But these bad things come with benefits. The predatory species help with pest control, burrowers and nest builders help maintain soil health, and the myriad of species that decompose dead organic matter stop us from being neck-deep in even more ugly material. Everyone, especially politicians and technocrats, must learn about soil biodiversity.

Approximately 23,000 beetle species have been formally described and named in Australia, but entomologists believe the yet-to-be-discovered species would bring the total to around 35,000 or even higher.

Soil biodiversity is important

In 1987 I went to the University of Zimbabwe on a Commonwealth Universities fellowship where for two years, I was a bushy-tailed post-doctoral scientist. It kick-started my career as an ecologist and I'll be forever grateful. It was the opportunity of a lifetime.

At that time, Zimbabwe was a vibrant place still buzzing with the euphoria of a newly minted country. Most of its infrastructure and economy were still functioning despite sanctions during the independence struggle. The friendly and accommodating locals all wore a wide grin.

A handsome young academic starts at his new job in Zimbabwe. You can almost smell the naivety of a 26-year-old Brit in mother Africa.

As a privilege of my scholarship, I became a member of the Tropical Soil Biology and Fertility (TSBF) program, a research group based in the Department of Biological Sciences but with a global remit. At various seminars and field visits across Africa, I was introduced to and helped by many senior soil biologists from all around the world. That exposure so early in my academic career was extremely valuable. It made me realise what it took to achieve academic success.

One individual, Professor Diana Wall, became a mentor and invited me to several workshops and conferences she organised on above- and below-ground biodiversity. Her pioneering work in soil ecology in the Antarctic is seminal even now, as was her leadership of the Natural Resource Ecology Laboratory at Colorado State University from 1992 to 2006. NREL has helped many thousands of youngsters on the way to productive ecological contributions partly because they all learned of the importance of soil biology.

After a long and distinguished career, Diana continues to promote soil biology as a critical missing link in our understanding of how the world works, especially for food security. In this, she's correct because, for some reason, humanity has no desire to understand what's going on beneath our feet.

Soil is the most important component of our food, as without it, we wouldn't have any. A small fraction of global food is produced in greenhouses with their roots in water or even air but 99% of the plants that grow on the planet have their roots in soil, including food crops and the livestock that feed on the pastures or are fed fodder crops in feedlots. It's all down to soil, but only a handful of specialists and some canny farmers know much about it.

We don't even know where to put it. For example, in New South Wales you can find soil in several government departments from water to planning to agriculture, environmental protection, and conservation. The already sparse expertise is split and separated around these various government institutions. Similarly, soil research is partitioned into its various components, so agronomists do their soil work while soil scientists grapple with soil physics, soil chemistry, and mapping. The ecologists have a crack at soil, too, typically with a focus on the soil animals and microbiome. These specialists are even fewer in number and must try to understand how organisms influence soil processes without knowing much about the myriad of soil organisms that make it happen.

Soil profile showing layers and plant root architecture. Miombo woodland, Zimbabwe.

Lack of focus on soils, particularly on soil biodiversity, is perplexing. The animals and microbes in soil, the soil biodiversity, supports the transfer of nutrients and water to plant roots, the storage of carbon, and is the foundation allowing plants to grow tall. It is the engine that keeps us alive.

When an engine is exploited but ignored, it breaks down sooner or later, and we're seeing this happening repeatedly throughout the world with a range of different consequences, none of which are pretty. Half of the world's soils have been impacted by human activity over the past 150 years, primarily through soil organic carbon (SOC) stock loss, impacting soil health and functionality. This soil degradation reduces nutrient provision to plant roots, compromises water retention, results in biodiversity loss, increases erosion vulnerability, and decreases crop yields, collectively impacting soil health intricately linked to food security, environmental stability, and overall sustainability. Current rates of erosion on agricultural land are estimated to be an order of magnitude higher than that of natural erosion or soil formation processes. Concepts like soil health that reflect the ability of soils to support plant growth do breakthrough occasionally but any metrics that describe such concepts are going in the wrong direction.

A handful of soil biologists around the place know these realities and have explained it in various forums. But their message is often forgotten or ignored. Even the farmers who understand the importance of soil and can tell you good soil from how it smells don't pass this on to the general public. In policy and public perception, there's no penetration at all, no understanding of the importance of soil in the public mind.

Even agricultural soil is full of biology

Soil is more important than climate

Arguably soil health is far more important than climate change to our long-term survival because without healthy soil there is not enough food for us to survive long enough to fix the climate. All a bit melodramatic perhaps, yet there is truth. Food insecurity is a destabilising risk that is with us right now. We wouldn't live very long if all soils became unhealthy. Yet, with some collective effort and the infamous technological agility of our species, we could survive most of the climate change impacts even if they were disruptive. But only if the soils are in good shape.

Although she's nearing the end of her career, Diana Wall is still active in promoting soil biodiversity as a critical component of our understanding.

> Geisen, S., Wall, D. H., & van der Putten, W. H. (2019) Challenges and Opportunities for Soil Biodiversity in the Anthropocene. Current Biology, 29(19), R1036-R1044.

In one of her recent papers, I found one of the reasons why soil biodiversity is considered such an esoteric subject with very little understanding by the general public. It's not that soils are hidden away or that most soil organisms are tiny microscopic critters that are hard to understand or that they only do their job because of their vast numbers. It was that the words used suggest that soil biodiversity is an object, like a koala or an elephant is an object.

Soil biodiversity is something they said; it's a thing. It's a component of something, perhaps a piston in an engine. You can see it, decide whether it's in good shape, and see if it needs oil or grease. It's a specific thing. By extension, the engine doesn't work if specific things are lost. If specific things become corroded or bent out of shape again, the engine is compromised or doesn't work at all. This idea of objectifying is how the conservation movement operates. They identify objects to save and protect.

I have to say that I was a little bit surprised. My recollections of Diana Wall's contributions and her intellect were that she's way smarter than that. She knows much more about ecosystems, but the communication was designed to get traction. The intent was to make information accessible to a wider audience. But in objectification, everything about why soil biodiversity matters is lost because ecology is not about objects. **It is all about processes**. And in my view that applies to biodiversity too. It is all about process, especially in soils.

Soil processes are important

Biological objects are present in soil and contribute to delivering processes, but the processes matter—decomposition, nutrient cycling, water balance. It is knowing and understanding resistance and resilience in these processes that is critical to how humanity is going to persist, even if it is at all.

I suspect that objectification is a mistake and a real problem, particularly when the soil biodiversity objects are so poorly known and have more yuck factor than charisma. Scientists are still unsure of soil organisms' types, numbers or distribution patterns. Very little is known about how to define the types, given that species concepts fall over for many soil organisms. What we do know is that soil biodiversity is incredibly diverse. Like rainforests, it is vulnerable to disturbances outside the norm, such as land clearing, for example, or grazing rangeland with livestock rather than bison.

Objectifying may make sense to get a message across. But by focusing on objects, we fall into the preservation of objects trap, and that's not what is needed. Global food security requires the maintenance of soil processes. And it doesn't matter which objects were involved as long as we can be sure that the processes themselves remain.

The message is this—understand the processes. Find out why soil processes are resilient and resistant for as long as it takes. We're talking about a requirement to produce food at 22 trillion calories a day for at least a hundred years. Our human problem is acute, and the only solution is to maintain soil processes long enough for humanity to solve its many other pressing issues.

Soil carbon and the biodiversity that goes with it are easily lost

Soil biodiversity is so rich that we cannot do this one object at a time. The soil equivalent of saving the koala does not exist. The curious thing is that we know all this already. Even the UN has reported on it.

25% of animal species

The core messages of *State of Knowledge of Soil Biodiversity*, a 2020 UN Food and Agriculture Organization report, are that soil is vital and that we know too little about the intricate workings of soil biology. The report is heavy stuff—quite literally, it is over 450 pages long—and I couldn't manage to read it all.

A quarter of all the animal species on Earth live beneath our feet. They scurry, feed and metabolise, moving nutrients around the soil. Plants take up these nutrients via their roots and build biomass called food.

Soils also store as much carbon as all plants above ground and will be part of the solution to the climate emergency.

The *Summary for Policymakers* is a little more accessible at 40 pages, which I can condense down for you into a few key points. In this list, the text from the summary report is in italics, and my shorthand version is in bold.

- *Soil organisms drive processes that produce food, purify soil and water, and preserve both human well-being and the health of the biosphere.* **Soil organisms do critical stuff**

- *Our current understanding of the role of soil organisms in plant growth and the transformation of pollutants has been harnessed to improve agricultural production and reclaim degraded soils.* **We already use soil organisms**

- *Laboratory and analytical advances in the past decade allow us to move beyond research on individual species to study whole communities of organisms, and hence develop new approaches to address food security and environmental protection.* **Innovation is here**

- *The essential contributions of soil organisms are threatened by soil-degrading practises. Policies that minimise soil degradation and protect soil biodiversity should be a component of biodiversity protection at all levels*. **Soil degradation is a huge problem**

These are standard, if impenetrable, suggestions that you would expect from the United Nations. Crudely translated, the message is that

humans are great, but maybe there are some issues that the masses don't know about.

It turns out that the key points in the report are supported by solid, logical sentiment from a thorough research base. A long list of references are presented to give credibility—over 140 pages of them in the main report. In casting a mindful sceptics eye across this work, we see a comprehensive review of evidence that covers most of what science knows is important about soil biodiversity.

The policymaker's summary advises championing soil biodiversity, knowing more about it, and making the rest of us more aware that soil is not just dirt. These are all noble objectives, yet there is always the nagging question of why, if dirt (sorry soil) is so important, does it barely get a mention in the real world?

While there is a clear scientific distinction between soil and dirt, this difference is not widely understood or appreciated by the general public. Improving soil literacy remains an important goal for environmental education and public awareness campaigns.

The way forward

After telling us about the problem, here is what the UN report says is 'the way forward' with my pithy version in bold.

- *Advocate for mainstreaming soil biodiversity into the sustainable development agenda, the Post-2020 biodiversity framework, the UN decade on ecosystem restoration, and all areas where soil biodiversity can contribute.* **Advocate**
- *Develop standard protocols and procedures for assessing soil biodiversity at different scales.* **Standardise assessments**
- *Promote the establishment of soil information and monitoring systems that include soil biodiversity as a key indicator of soil health.* **Share information**
- *Strengthen knowledge of the different soil groups forming soil biodiversity (i.e., microbes, micro, meso, macro, and megafauna); improve knowledge (including local or traditional) of the soil microbiome.* **Learn more**
- *Establish a global capacity building programme for the use and management of soil biodiversity and the Global Soil Biodiversity Observatory.* **Build capacity**

Suddenly, the essential urgency is lost in the technical mumbo jumbo. The way forward sounds more like a research agenda than awareness or practical action. Odd for an issue that impacts '*human well-being and the health of the biosphere'.* It also sounds very familiar to me. It is the same set of ideas and actions that Diana Wall and her TSBF colleagues introduced to me back in the 1980's and we wrote most of it down.

- Smith, Pete, Olof Andren, Lijbert Brussaard, Mark Dangerfield, Klemens Ekschmitt, Patrick Lavelle, and Kevin Tate. "Soil biota and global change at the ecosystem level: describing soil biota in mathematical models." *Global Change Biology* 4, no. 7 (1998): 773-784.

- Young, I. M., Blanchart, E., Chenu, C., Dangerfield, M., Fragoso, C., Grimaldi, M., ... & Monrozier, L. J. (1998). The interaction of soil biota and soil structure under global change. *Global Change Biology*, *4*(7), 703-712.

- Wolters, V., Silver, W. L., Bignell, D. E., Coleman, D. C., Lavelle, P., Van Der Putten, W. H., ... & Van Veen, J. A. (2000). Effects of Global Changes on Above-and Belowground Biodiversity in Terrestrial Ecosystems: Implications for Ecosystem Functioning: We identify the basic types of interaction between vascular plants and soil biota; describe the sensitivity of each type to changes in species composition; and, within this framework, evaluate the potential consequences of global change drivers on ecosystem processes. *Bioscience*, *50*(12), 1089-1098.

- Hooper, D. U., Bignell, D. E., Brown, V. K., Brussard, L., Dangerfield, J. M., Wall, D. H., ... & Wolters, V. (2000). Interactions between Aboveground and Belowground Biodiversity in Terrestrial Ecosystems: Patterns, Mechanisms, and Feedbacks: We assess the evidence for correlation between aboveground and belowground diversity and conclude that a variety of mechanisms could lead to positive, negative, or no relationship—depending on the strength and type of interactions among species. *Bioscience, 50*(12), 1049-1061.

The mindful sceptic knows that humanity is in lockstep with ecological processes, and we should all have some knowledge of this fact. Soil biodiversity should be in everyone's face to make it everyone's problem to resolve. If we leave this all to the ivory towers, nothing meaningful will get done for years, realistically decades. By then, it will be too late.

What we know right now is that **the biology beneath our feet is critical to the future of humanity**. We know what impacts this biodiversity and what it does, so we also know what not to do to land and what to do to land to keep biodiversity in good shape. We have to get on with it, especially on farms.

Farming is the global issue

Food production is essential for global sustenance, shaping landscapes, soils, and rural cultures. With over half the world's population residing in urban areas, most people depend on complex supply chains to meet their nutritional needs, as most food is cultivated in rural fields.

Agriculture, including current practices and historical land clearing, contributes significantly to greenhouse gas emissions—accounting for approximately one-third of the total. This environmental impact persists and may even increase as food production must continue to meet growing demand.

While necessary for food production, agricultural landscapes often create environments inhospitable to biodiversity. As natural habitats are reduced and fragmented, species loss becomes inevitable. Modern farming methods have made rural areas increasingly challenging for various life forms, particularly crucial groups such as insects and microorganisms.

Our current food systems, characterized by industrial-scale intensification, are precariously balanced. While they meet production demands through heavy reliance on energy inputs and chemical additives, these systems are vulnerable to foreseeable and unexpected disruptions. The dependency on these inputs creates a fragile system that risks collapse if the supply chain is interrupted.

But it is not all bad news. Thanks to proven, science-based ways to create more efficient, less polluting and more resilient agricultural landscapes with less mining of soil, our prescience tells us there are lots of actions we should take, for example

- keep vegetation cover on the soil at all times and reduce tillage to a minimum

- protect patches of remnant native vegetation—habitat for animals and helps draw down and store carbon from the atmosphere

- retain water in the landscape by slowing the overland flow and creating healthy farm dams which can provide higher-quality drinking water for livestock, improve farm productivity and create wildlife habitat

- plant "shelterbelts", strips of woody vegetation that shelter livestock from wind and sun, provide wildlife habitat (when well designed and managed), and prevent moisture loss from soil

- control weeds and pests

- shorten supply chains wherever possible
- reduce waste so that the majority of food produced gets eaten

The most important thing is maintaining as many of these actions as possible and more **simultaneously** because no single solution works. It must be integrated.

There is nothing in future food security that we cannot see coming, and we already have most of the evidence to deliver and justify actions to mitigate risk. The stark truth is that we know the problem presented by biodiversity loss in soils—fragility, risk, and unprecedented demand. We know the solutions—integrate multiple actions that maintain healthy soil, reduce waste, and use more ecology to reduce inputs. We have to get onto it.

Insectageddon is a huge deal. It is a symptom of the intensification of agriculture and the scale of human exploitation of nature. Nothing should trivialise this loss. When the bees go, we are in deep trouble. When the decomposers are compromised, the flow of nutrients to crops is too. We need the bugs.

But the real problem is that we don't know we are in nature. We still believe that we are somehow separate and must subdue the Earth. Our place is somehow different from all other species; we have dominion.

We have built our economies as though they are separate and autonomous despite acquiring all the raw materials and energy from nature, and then using the land and oceans to dump our waste. All of this is crazy and foolish.

Humans are a species like any other, dependent on finite resources on a finite planet. We might be smart enough to bend nature to extremes, but until we recognise our dependence, those smarts will wreck the future. An insect apocalypse is a harbinger.

Key points

- Insects and other invertebrates are experiencing alarming declines, with far-reaching consequences for ecosystems and their services, such as pollination, pest control, and decomposition.

- Soil biodiversity is essential for food production, nutrient cycling, and ecosystem health but is often overlooked and poorly understood. For long-term human survival, soil health should be considered as important as, if not more important than, climate change.

- Focusing on soil biodiversity as a process rather than a collection of objects is crucial for maintaining the critical functions that soil organisms facilitate. Efforts should be directed towards preserving these processes to ensure soil's long-term health and productivity.

- Agriculture is a significant driver of biodiversity loss, but science-based solutions exist to create more sustainable, resilient, and efficient agricultural landscapes. Implementing these solutions, such as reducing tillage, protecting native vegetation, and reducing waste, is essential for addressing the biodiversity crisis and ensuring food security.

In the next chapter

It is time to confront the undeniable truth.

Chapter Four

The reason for biodiversity loss

There is one reason for biodiversity loss currently making up the sixth mass extinction, and it is beyond time that it was owned. Here are the basics from an unusual angle.

Human actions

When conservationists lament the loss of habitat and the species that live in it, they are angry and sad. Emotions are triggered and often result in railing at the system that cleared the habitat, polluted the rivers and oceans and hunted the wildlife. It is much harder to focus on the real reason for biodiversity loss. Here it is.

Estimates suggest that 96% of the mammal biomass on earth today—that's the weight of the mammals that we can see with our eyes—is made up of humans and livestock, the animals that human beings manage for our benefit. I'll just say that number again... **ninety-six per cent**.

Almost all the mammal biomass comprises cattle, sheep, pigs, goats, horses, companion animals, and the occasional alpaca. Here again is what that looks like.

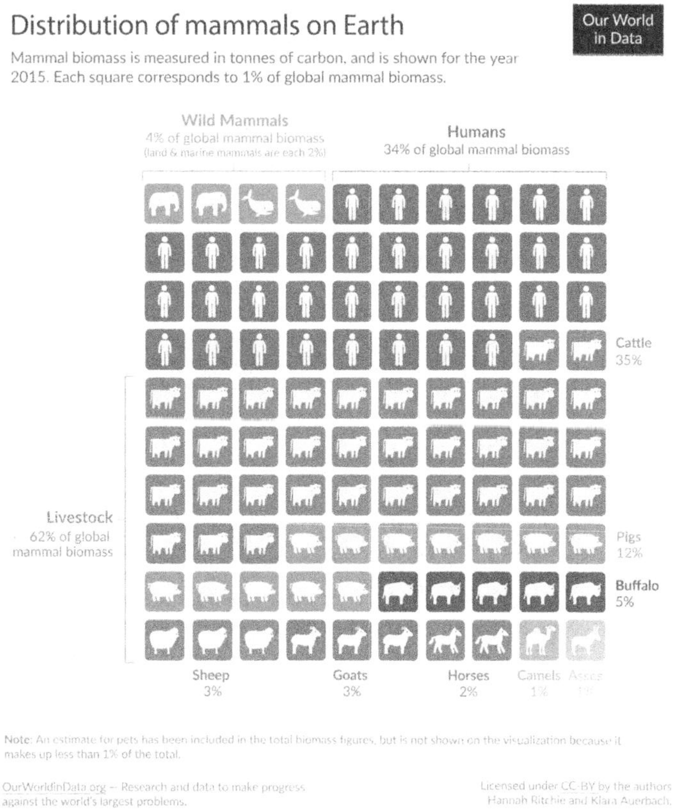

And just in case, I will say it again, 96%.

That means just 4% of the biomass of all mammals is available to distribute among the mammal species not used as human livestock. And marine mammals account for half of that 4%. In 2008, the International Union for Conservation of Nature (IUCN) completed a five-year Global Mammal Assessment for its IUCN Red List that counted 5,488 mammal species. Recall that biomass is the physicality of the organism. The mass of an object is a measure of the object's inertial property or the amount of matter it contains and a measure of the force exerted on the object by gravity. So, of all of the mass in land mammals, only 2% is available to be distributed among thousands of wild mammal species. So here is the truth.

Human actions are the reason for biodiversity loss. It is true we are to blame for the sixth mass extinction. And this should not be a surprise given there are now 8 billion of us alive at the same time—put another way, 7% of all the *Homo sapiens* that have ever lived are living right now.

As of 2021, there were approximately 14 million cattle in U.S. feedlots at any given time. The total one-time capacity of U.S. feedlots is around 17 million head

Some basic ecology is needed here. The sun shines and delivers energy to the Earth. Some of that energy is captured by plants and converted into biomass that animals utilise. The microbes and fungi are around too to help with the nutrient transfers, and the whole ecosystem has a maximum amount of energy transfer determined by the location, the season and the climate.

There is maximum biomass that the planet can deliver, and what's happened is that humans have appropriated, directly or indirectly through the animals that they consume, a large proportion of that biomass. Agriculture and, in particular, livestock husbandry has resulted in a huge proportion of primary production going into that 96%. In some places, the intensification has increased the biomass potential by adding extra nutrients and energy. However you look at it, humans have gone from being responsible for less than 3% of the mammal biomass in pre-industrial revolution days to 96% today.

Animals respire. That means they use energy from their environment in the form of other organisms to survive, grow, and reproduce. The energy conversion rate isn't great, and mammals with warm bodies need to eat a lot, especially plant eaters. And it's the herbivores that humans use. We are not fond of carnivores; anyway, there are relatively few of them due to that energy conversion problem.

Basic thermodynamics and ecology mean that at least 50% of the planet's habitable land area is used for agriculture, and 77% of that, some 40 million km2, is for livestock rearing. In other words, humans have converted vast amounts of land to either graze or grow feed for our livestock. The biomass of land mammals and the land needed to support them have impacted the biodiversity of not just the other mammal species but also the plants and invertebrate animals.

Since their energy sources have been stolen, many mammal species will inevitably be under great pressure to survive. As will so many other species deprived of their energy source. Somehow I don't understand what is hard about this basic thermodynamics. We can't have our steak and eat it in this case. You can't appropriate all the biomass and expect to keep all the other animals you like.

As an ecologist, I find this logic so basic and ingrained that it's a no-brainer. We shouldn't be surprised about any of the biodiversity loss we see worldwide. The appropriation of energy and biomass is sufficient to make it happen. Add habitat change, climate issues, and altering habitats through pollution and other externalities that businesses are keen on delivering to the world. It just makes biodiversity loss inevitable—the **reason for biodiversity loss is that humans have to eat.**

Combining all species, the total zebra population in Africa is likely between 510,000 and 760,000 individuals.

Farming our only planet

It's also hard to dispute the numbers. Whether it's 96% or 79% is not the issue. Load Google Earth and pan across any of the continents at the height of around 1,000m; you can see that number in the bottom right-hand corner of the screen and maybe start at your home town. What you will see is a landscape that has been altered by human activity. There will be a mosaic of fields, maybe some forests or grasslands, but the evidence of humans altering the landscape for food is everywhere. There are very few countries with more than 10 or 15 percent of their land area that's in anything other than agricultural production. There are a few reserves here and there and a few patches of vegetation here and there, but essentially, the landscape has been converted into one massive farm for the benefit of humanity.

There will always be a dispute over the numbers, dare we say denial, but the imagery is there to be seen, and if you don't believe the satellite imagery, drive through the countryside. When driving, look beyond the fringing vegetation on the roadside and into the fields. And you'll see that they go for miles. We have converted nature into landscapes; there's no question about it.

A legacy of global land use change to accommodate crops and livestock—the real reason for biodiversity loss—is a hard one to take. The reality and the magnitude of this change to the planet, our only planet, isolated in this tiny corner of the universe. The only place that we have, the only place we'll ever have, is being ripped from underneath us by our actions.

That is what it is. The only hope is that we are smart. Having figured out all these ways to appropriate resources, maybe we're smart enough to realise what we're doing and turn it around. There's a small window of opportunity over the next 50 to 100 years, but we have to start now. Or that window will shut.

Just in case you needed more.

On 11 March 2020, WHO declared the outbreak of COVID-19 a pandemic. Two months later, the first version of this chapter was written. As I write these edits, it is 6 June 2023. The time lapsed is 1,182 days, a little over three years. In the period of time lapsed, there are now 235,948,199 more people to feed, clothe and house because, with or without lockdowns, the human population is growing at a net rate of roughly 8,000 humans an hour.

When these new people become adults, they will go about their business, just as all the rest of us are doing. And we can estimate the consequence of that activity in greenhouse gas emissions. At the average global rate of emissions per capita in 2021—4.7 tCO_2e per annum—these people will emit roughly 1,109 million tCO_2e every year throughout their adult lives—the same as the total emissions in 2016 of the UK, Italy and France combined. Think about the reason for biodiversity loss again.

It's not the cows; it's the people.

Key Points

- Human actions, primarily the conversion of land for agriculture and livestock rearing, are the reason for the ongoing sixth mass extinction. 96% of mammal biomass on Earth consists of humans and livestock, leaving only 4% for all other wild mammal species combined.

- Animals require significant energy from their environment to survive, and the Earth has a maximum amount of biomass it can support. Humans have appropriated a large proportion of this biomass, severely impacting the biodiversity of mammals, plants, and invertebrates.

- The reason for biodiversity loss is human consumption and the need to feed a growing population. Denying this fact is futile, as evidence of the extensive conversion of landscapes for human benefit is readily visible.

- The growing human population and their consumption patterns, not just livestock, are the primary drivers of biodiversity loss. While the situation is dire, there is still a small window of opportunity to turn things around. However, action must be taken now.

In the next chapter

In a world where the relentless march of human progress leaves a trail of extinct species in its wake, there is an unsettling truth that extinction is not just a consequence of our actions but a stark reminder of our own fragility.

Chapter Five

Extinction is forever

Extinction is forever, and there is more of it going on just now than normal, making what causes extinction a crucial question.

All organisms evolve

What causes extinction, let's say, of the dodo, a flightless bird endemic to the island of Mauritius in the Indian Ocean? Nobody knows what the huge pigeon-like creature looked like, given the last widely accepted sighting of a dodo was in 1662, and previous descriptions and drawings were hard to verify. Many were made by illustrators who had never seen the animal.

Dodos had not eyeballed humans either and had no fear of them, but there is no evidence that sailors plundering populations for provisions sent them to extinction. Nor was it likely that the resident humans on the island, which never exceeded 50 people in the 17th century, ate them to death. People cleared much of the dodo's forest habitat and introduced alien animals to the island, including dogs, pigs, cats, rats, and crab-eating macaques, which plundered dodo nests and competed for the limited food resources. Nobody knows precisely what happened.

What we do know is that after Dutch sailors first reported the dodo in 1598, the species was probably extinct a century later. Indeed, when the Dutch left Mauritius in 1710, the dodo and most of the large terrestrial vertebrates on the island were gone.

So, what causes extinction? There is a one-word answer, but before we reveal that, let's look at evolution and when extinction happens.

Artistic depictions of the dodo have varied over time, and early representations may not have been entirely accurate.

The dodo was well suited to pre-human life in Mauritius. We know this because it existed as a distinct species to be seen by the intrepid explorers in their wooden ships from far away. It had evolved into its shape, behaviours and habitat over thousands of generations from a slightly different-looking ancestor.

The ancestors of the dodo each varied in anatomical, physiological and behavioural attributes generated through genetic mutation and drift. Some of these attributes were a benefit, and some were a hindrance. Benefits we call an adaptation to the environmental conditions in which that organism finds itself that gives the individual a better chance of survival, growth, and reproduction. Indeed, a benefit in evolutionary terms only occurs if the individual successfully reproduces. How much help future generations get from a mutation depends on the environmental conditions and if those conditions are stable relative to the benefits the new attribute brings.

The dodo was flightless. It had stubby wings, too small to lift the bulky animal off the ground. If the ability to fly offered no advantage to survival, growth or reproduction over not flying, then a mutation that made wings smaller had as good a chance of propagation as fully functional wings. Meanwhile, if a large body and big beak gave foraging advantages and no penalty for not flying, before long, dodos without expensive wings and flight muscles had the advantage. Biology sees no need to build and maintain unnecessary structures. Individuals carrying the genes for big wings were at a disadvantage, and natural selection weeded them out. It is cruel like that.

Here is another example. Suppose a tree species evolves a mutation that allows seedlings to survive longer when the soil is dry. This might be a considerable benefit if the climate is warming and the soil drying out faster between rain events. A mutation for more robust seedlings allows for regeneration and buys some time for future generations of the tree to track habitat change as warmer, drier conditions spread across the landscape. If the climate changes to colder and wetter, the mutation may have no effect or even be a disadvantage. This is the thing. The mutations that offer an advantage will propagate until they don't.

When does extinction occur?

The last lone rhino without a mate might persist for a few decades, and when it dies, the species is extinct. Extinction happens when the last individuals of a species can no longer reproduce unaided.

Failure to reproduce can occur due to various reasons. One cause is death before reproduction, which happens when an organism fails to survive because the conditions were too severe or there was a lack of food, water, or nutrients, resulting from factors ranging from competition to a tsunami. Another reason is insufficient food or nutrients, which can prevent an organism from growing to reproductive size or channeling resources into gametes, or both. Lastly, the absence of necessary gametes can also lead to reproductive failure, which can happen for a range of reasons, from the failure to find a mate to a shortage of pollinators.

Failure to reproduce for the last remaining individuals is the point of extinction. And failure to reproduce is about the survival, growth and reproduction of individual organisms. Enough individuals must achieve successful reproduction for the species to persist. This basic premise applies to everything from microbes to marsupials.

Extinction happens when the ability of the organism to gather sufficient resources is compromised to the point of failure to reproduce in all the remaining individuals.

What causes extinction in the specific sense is the failure of all remaining individuals of a given species to gather enough resources to survive, grow and reproduce—the specific failure depends on the species and the circumstances. Extinction is much more likely when the species is rare, has a restricted distribution, or both. The dodo on Mauritius had a restricted distribution and became rare when humans became unfamiliar predators and altered the dodo's habitat.

Extinction is also likely when environmental conditions change sufficiently to compromise the ability of individuals to reproduce. Should environmental conditions change dramatically from what is usual, for example, a human being comes along with an axe and chops a bunch of trees down, this is a disturbance outside the norm in terms of extent and intensity. Many organisms are pushed outside of their comfort zone and struggle in new conditions. Any remaining **individuals** have only three options.

1. Hunker down in whatever safe place they can find and hope the disturbance disappears and normality returns.

2. Move to the nearest patch of intact habitat; easy for an eagle, hard for an orchid or a dodo.

3. Die.

Adaptation through genetics happens quite slowly, and extinction occurs when the environment no longer suits the evolved adaptations. We are close to the one-word answer to what causes extinction. Just stringing it out a little longer, look at this image from Western Australia that shows intact eucalyptus forest and forest cleared for agriculture. Now, it doesn't take a genius to work out that the conditions on one side are very different to those on the other. Many of the organisms that exist in the three-dimensional conditions of the forest—places to hide, food to eat, shade from the sun—would struggle to find adequate conditions in the agricultural field.

Western Australia typically produces around 6-10 million tonnes of wheat annually. Average yields can range from 1.5 to 2.5 tonnes per hectare, depending on seasonal conditions.

What causes extinction?

The one-word answer is **disturbance**. Specifically, a disturbance that alters conditions beyond the tolerance of the last remaining individuals of a species sufficient to prevent them from reproducing.

Right now, disturbance is everywhere. To say that we're in the Anthropocene amid the Sixth Mass Extinction event is not a surprise if you look at how the globe now looks compared to what it did just a few hundred years ago. In the last two hundred years, humans have harnessed fossil fuels through machinery to transform the landscape radically. The vegetation is cleared and managed, and the organisms familiar with a slow, subtle change from predictable, natural disturbance can no longer cope. The changes that humans have brought about are radical, intense and acute.

People can sound incredulous and surprised by this reality, but it shouldn't be surprising. If we disturb such a large proportion of the landscape with axes, chains, fire, fertlizers and tractors, then we know that we will lose biological elements of that landscape. Any surprise we have at biodiversity loss is misplaced—it shouldn't be a surprise at all. Biodiversity loss through local and global extinction is an inevitable consequence of human beings acquiring resources and making humans the priority. As we channel nature to our purposes, humans disturb almost everything.

Disturbance causes extinction

Before we get all high and mighty or anthropocentric on disturbance, recall it has happened at a global scale many times. Human-induced mass extinction is at least the sixth big extinction event in the fossil record, along with evidence for numerous small global events. Most of them are attributed to climate changes, especially ocean warming, and even the famous meteorite strike creating the boundary between the Cretaceous and Tertiary approximately 66 million years ago that took out the dinosaurs induced a climate disturbance. Most of those creatures didn't die in the blast but starved when the dust in the atmosphere shaded the sun and messed with the plants. Disturbance is what causes extinction, and disturbances happen all the time.

The current human one is not unique. Nor is it immoral because it is inevitable; it is simply a consequence of adaptation... human adaptation.

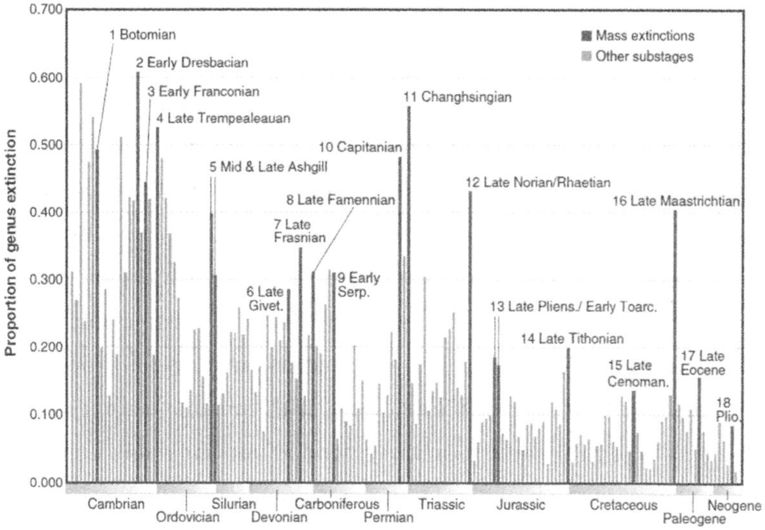

More than one extinction event recorded in the fossil record. By Marshall, Charles R. - "Forty Years Later: The Status of the 'Big Five' Mass Extinctions." Cambridge Prisms: Extinction 1 (2023): e5. doi:10.1017/ext.2022.4., CC BY 4.0, https://commons.wikimedia.org/w/index.php?curid=131040385

Disturbance is what causes extinction, and right now, the planet is disturbed by human activity, specifically activity in the last 200 years.

Extinction can be sad for us. It is tempting to focus on losses as preventable events and a bad thing with a cause and someone to blame. But if disturbance is what causes extinction, then it is going to happen because of what *Homo sapiens*, the species, has become. Not only are we swallowers of exogenous energy and able to bend the environment to our own needs, but we are also abundant.

Our challenge is not extinction but to cope with the disturbance consequences of rapid population growth. And when we say rapid, we mean rapid to levels that, historically, not even Thomas Malthus would have believed possible. But here we are at 8 billion, increasing by 8,000 extra people every hour. With this many people, disturbance is inevitable.

So we shouldn't be surprised at extinction. If 50 people on a 2,000 km^2 island and a few visitors in wooden boats can push a flightless bird to extinction, what are 8 billion humans with container ships going to do? It is time to become more pragmatic about our rhetoric around loss. Accept it for what it is. Then, we will make some very tough but important choices about what types of biodiversity we want to keep and, therefore, what effort it will take.

So don't be surprised when species go extinct. Be sad, disappointed, and somewhat ashamed that we haven't been able to fix this problem or manage our use of the planet more effectively. However, we must recognize that humanity faces a monumental challenge with both pessimistic and optimistic outcomes. Extinction is a risk that faces *Homo sapiens*, too. In the end, the question that we have to answer is not how much biodiversity loss is happening but how much disturbance and extinction can we tolerate before it starts to impact our likelihood of survival.

Biodiversity reduction

Books often have gold nuggets hidden away in them, somewhere in the slower middle chapters where the reader is no longer flush with excitement for the journey or anticipating the punchline. This is one for this book. **Biodiversity loss is not just about extinction**; reduction is important, too. What biodiversity does to deliver and maintain ecological processes in nature can decline without extinction because that means all individuals of that type are lost. There can be local species loss, and yet the species still persists somewhere else.

We focus on loss, especially extinction, when it's reduction that matters.

Key Points

- Extinction occurs when the last individuals of a species can no longer reproduce unaided, which can happen due to various reasons such as death before reproduction, insufficient resources, or the absence of gametes.

- Disturbance is the primary cause of extinction. It alters conditions beyond the tolerance of a species' last remaining individuals, preventing them from reproducing. Human activities have caused widespread disturbance, leading to the sixth mass extinction event.

- Disturbance and extinction are not unique to human activities, as the fossil record shows evidence of multiple extinction events caused by various factors. However, the current human-induced disturbance is occurring at an unprecedented scale and pace.

- With a human population of 8 billion and growing, disturbance and extinction are inevitable. Instead of focusing solely on preventing extinction, we should consider the level of disturbance and extinction we can tolerate before it starts to impact our own survival.

In the next chapter

Why do we still set international biodiversity targets when we know that relentless expansion and dominion over nature are the root causes of the crisis?

Chapter Six

Why international biodiversity targets fail

Humanity has taken dominion and then wonders why we fail to meet international biodiversity targets.

International biodiversity targets

The UN biodiversity initiatives, which were the first international biodiversity targets, began in Rio in 1992, and they were bold. Heads of government joined their environment ministers in committing to support and save the diversity of life, which at the time was a new description of nature. As countries signed the Convention on Biological Diversity (CBD), there was hope that species loss and habitat destruction could slow or even halt. Then it all went belly up.

Nearly 20 years later, in October 2010, new international biodiversity targets were set. At the 10th Conference of the Parties meeting in Nagoya, Aichi Prefecture, Japan, a revised and updated *Strategic Plan for*

Biodiversity, including the Aichi Biodiversity Targets, for the 2011-2020 period was declared with the following fanfare...

> *an overarching framework on biodiversity, not only for the biodiversity-related conventions, but for the entire United Nations system and all other partners engaged in biodiversity management and policy development.*
>
> UN Aichi Biodiversity Targets

A new set of international biodiversity targets was set. This was a tacit admission that, in two decades, the original convention had done little to combat biodiversity loss and habitat degradation. Instead, it announced another decade to achieve new bold goals.

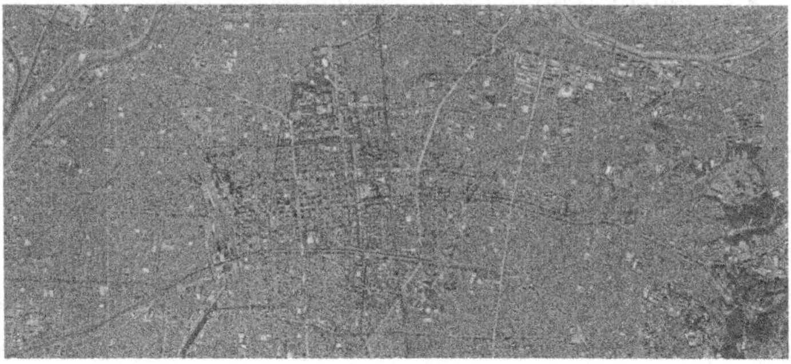

Nagoya, Aichi Prefecture, Japan has a population of approximately 2.3 million people. It is the fourth-largest city in Japan by population.

Over a decade on from Aichi and reports are out now that suggest almost all of those Aichi international biodiversity targets were not met. There was a movement towards some of them, in others, no change or further decline. This is rocket fuel for those who believe that the United

Nations and other global collaborations are just a waste of time. Talk fests for the privileged to travel around the world to posh hotels and conference centres to have convoluted conversations that mean nothing.

The reality, of course, is that in the decade since the Aichi targets were set, the human population has increased by 800 million, the global economy has increased by 24%, a US$17,600 billion increase in global GDP, and food production has risen by 15%, including cereal production up by 350 million tonnes, whilst humans and their livestock take up 96% of the mammal biomass.

Oh yes, and the climate has changed dramatically in large parts of the world. We now see fires in places that haven't seen fire in historical times. We now see wetlands drying not due to draining the catchments but due to changes in weather patterns. It is no surprise that the Aichi biodiversity targets were not reached. And that the best we could hope for was a movement towards them.

International biodiversity targets or not?

All this begs the question of whether we should even have targets. Whether presenting desires for biodiversity outcomes in international forums actually makes any difference at all. The logic for why we should goes something like this.

Signing up for an obligation for international biodiversity targets will force individual countries to take legislative action themselves in order to make things happen within their jurisdictions. A peer pressure agreement. Some countries have taken commitments seriously, some have used creative accounting to demonstrate some progress, and others have done nothing.

Accounting is one of the biggest problems with this whole story. How do we account for something that is as diffuse, nebulous and hard to define as biodiversity? It's easy to account for dollars. They exist as a currency to exchange for goods and services and can be converted to all sorts of other currencies, including energy. There are standard practices for understanding how to calculate debt and opportunity down the track. Biodiversity accounting is not so easy.

Biodiversity is a combination of the composition, function, and structure of assemblages of animals and plants. It's a complex set of ideas. There's no single currency to describe or measure its elements and characteristics. It can't be defined by rare species or the total number of species because it has to include what the species do, especially in creating ecosystem services, the performance of individual processes. It's a combination of all of these things. So, we have no obvious accounting tool with which to make measurements against targets. There is even a question mark over whether or not an accounting target is worth setting.

However, biodiversity is critical to our future. And in this instance, 'our' means individuals, societies, and humanity. All that biology wrapped up in species, assemblages and ecosystems is the engine of our food ecology and crucial for provisioning and regulating ecosystem services. It makes sense to want to look after biodiversity. Indeed, humanity would be crazy not to. So why are we failing?

Why we fail to meet international biodiversity targets

Take a look at this pair of images from Google Earth of central NSW, Australia which has a population density of 8.6 people per km^2, less really in this area as two-thirds of the population live in Sydney.

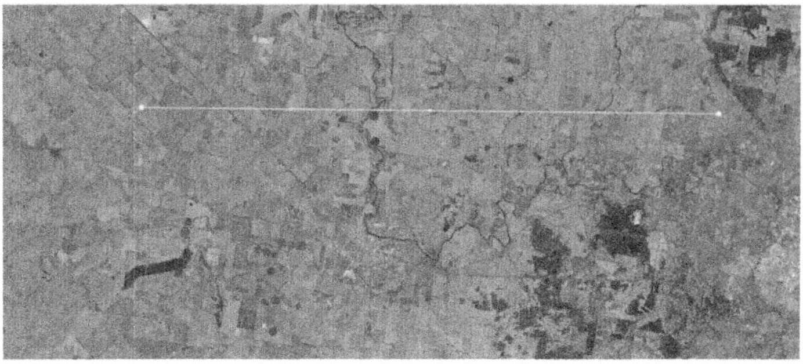

Google Earth image of central NSW near Dubbo captured at a 6,000m height. The yellow line represents 50km.

Close-up of the same location, this time from 1,000m and where the yellow line is 5km.

Here is another image pair slightly greener on a similar scale, this time for central England, with a population density of 281 per km^2.

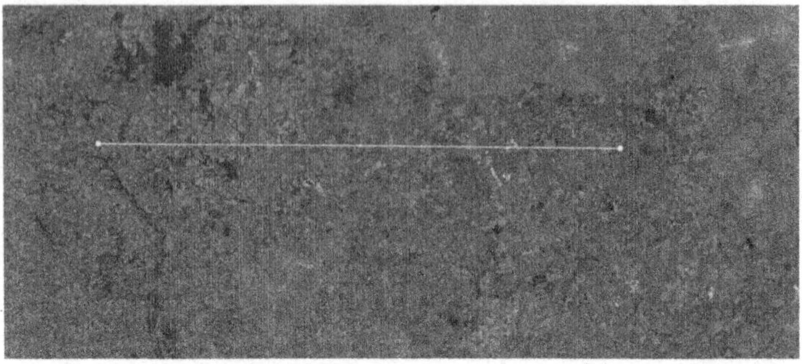

Google Earth image of central England south of Birmingham captured at a 6,000m height. The yellow line represents 50km.

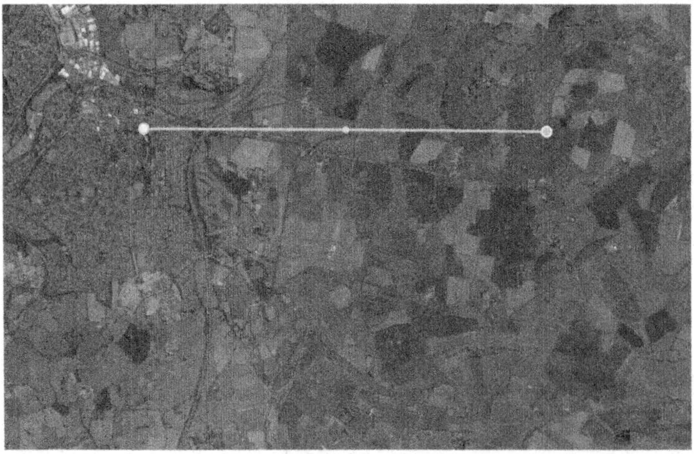

Close-up of the same location, this time from 1,000m and where the yellow line is 5km.

When you look at these images, the reason we fail to meet international biodiversity targets stares back at you. People have radically altered the face of the planet to create landscapes of agricultural production that we must have or we starve. It is very hard indeed to meet biodiversity targets when humans are not just occupying but altering just about every square meter of the landscape.

There will be complaints and arguments that we fail to meet targets because of a lack of will, too much rhetoric, and not enough commitment. Perhaps there is no hope for resolution given the vested interests, greed, and lack of empathy between nations and decision-makers, all wrapped up in some complex but powerful geopolitics. No doubt, these factors can vacuum up all the emotional energy. But the real reason we don't meet the targets is that they are too hard for a global population of people this large, with this big a footprint and a blind eye to the reality.

Take another look at those images. They are truly revealing.

The Leader's Pledge for Nature

People are nervous. No surprise, thanks to war, a pandemic, climate uncertainty and, where I live at least, some horrifying weather. Also, as a Pew Research Centre survey discovered, democracy is in strife, the social fabric is worn, and faith in institutions has faltered. All rather depressing even before we get to the environment that has been in decline for decades and seems no longer able to absorb human excesses.

Some good news is a boon. So here is some from a surprising source, the *Leaders' Pledge for Nature*. The pledge is an improvement on past commitments for targets. It is closer to an admission of the real problem—massive population expansion from the energy and materials subsidy from fossil fuels—with some high-level suggestions that would improve matters.

Launched before the UN summit on biodiversity in 2020, the pledge departed from decades of diplomatic rhetoric. In their opening remarks, the leaders say

We, political leaders participating in the United Nations Summit on Biodiversity, representing 64 countries from all regions and the European Union, have come together today, on 28 September 2020, ahead of the Summit to send a united signal to step up global ambition for biodiversity and to commit to matching our collective ambition for nature, climate and people with the scale of the crisis at hand.

Well, not so earth-shattering, perhaps. We have seen bold commitments before. Remember, we are talking about the 15th meeting of the Conference of the Parties to the Convention on Biological Diversity (CBD COP 15), where we get "*in order to highlight the urgency of action at the highest levels in support of a post-2020 global biodiversity framework that contributes to the 2030 Agenda for Sustainable Development and places the global community on a path towards realising the 2050 Vision for Biodiversity.*" I am not sure what happened at the other 14 COPs that go back to the original Convention on Biodiversity signed in Rio in 1992, except that biodiversity loss continues. Leaders still don't understand the importance of soil and soil biodiversity. But we are talking good news, and the new pledge seems slightly different and optimistic. Here are some of the better parts.

The pledge talks about **integration**. It mentions the need for **finance from the private sector** and sees **the problem as everyone's** rather than blaming one industry or another. After a prelude that emphasises how nature underpins all human activity, we can paraphrase the ten keys of the leaders' pledge for nature as

1. Make any recovery from COVID-19 green

2. Develop and deliver a transformational post-2020 global biodi-

versity framework

3. End silo thinking

4. Transition to sustainable food systems that meet people's needs while remaining within planetary boundaries

5. Raise ambition in climate policies

6. End environmental crimes

7. Mainstream biodiversity into policies

8. Address health and environmental sustainability in an integrated fashion

9. Strengthen financial means of implementation with resources from all sources, public and private

10. The design and implementation of policy will be science-based

Most of this makes good sense, the rhetoric factor notwithstanding. Green (1), framework (2) and climate policy (5) are standard claims, the same old, same old, right out of the technobabble phrasebook. Ending silo thinking (3) and mainstreaming biodiversity in science-based policy (7, 10) are excellent ideas so long as biodiversity is not a pseudonym for conservation, given the tendency to confuse essential terms. In our experience, ending silos and mainstreaming into policy will be tough but worth trying.

Transition to sustainable food systems (4) is a mindful sceptics mantra for sustainable food, so we love that one. Health and sustainability (8) we call sustainable diet, and is a great place to start.

All societies should be looking to reduce criminal activity (6), and we recommend that this extends to punishing the externalities that the capitalist system is so good at avoiding. Finally, we love the idea of innovative finance (9) so long as we heed Einstein and don't let the same ideas that got us into this mess masquarade as solutions.

There is much to like in the pledge, even at this high level.

The Serengeti-Mara ecosystem is home to approximately 1.5 to 2 million wildebeest. Perhaps the pledges might save some of them.

It is a pity that not all countries signed the leaders' pledge for nature. In Australia, the Morrison government refused to sign. They claimed that a global pledge endorsed by 64 countries committing them to reverse biodiversity loss because it was '*inconsistent with Australia's policies*', including a greater ambition to reduce greenhouse gas pollution and reach net zero emissions by 2050. What sort of excuse is that from the country with the second-highest per capita GHG emissions? A poor one.

Apart from Australia, other countries that didn't sign the pledge include the United States, Brazil, China, Russia and India. These four countries represent 3.5 billion people, some 45% of the global total. A commitment by a little over half the people dilutes the collective ambition. Only some people elect or are led by leaders up for the challenge. Indeed, it makes a race to the bottom more likely than ever.

Then there is the reality of why biodiversity loss is an issue in the first place—given that we set dominion over nature as our excuse for the pursuit of excess and a failure to pull up our biological instincts to make more means that we don't meet any of the targets we set. They are a sop.

Kangaroos are a natural and important part of Australian ecosystems, but their interactions with human activities can lead to conflicts in certain situations.

My pessimism must be infuriating, and I apologise. The leaders pledge for nature is an improvement on past commitments. It is closer to an admission of the real problem—massive population expansion from the energy and materials subsidy from fossil fuels—with some high-level suggestions that would improve matters.

Global diplomacy is indeed slow. In the case of biodiversity, it has taken 30 years to get to this level of understanding that acknowledges integration, that it's everyone's problem and that we need to throw money at solutions. But slow progress is better than being stuck or in denial.

We hope that the leaders pledge for nature triggers a strategy, not just a set of aspirations. What has to happen is that youngsters, perhaps readers of this chapter, take this pledge and give it a Red Bull. Inject some youthful momentum into it. The **fearless enthusiasm of youth** would be our 11th and master key, the one to open all the doors.

Key Points

- Despite the establishment of international biodiversity targets, such as the Aichi Biodiversity Targets, most were not met due to the significant increase in human population, global economy, food production, and climate change.

- The primary reason for failing to meet biodiversity targets is the extensive alteration of landscapes for agricultural production to support the growing human population. Given the current global population size and its environmental footprint, the targets are too difficult to achieve.

- Accounting for and measuring biodiversity is challenging, as it is a complex combination of the composition, function, and structure of animal and plant assemblages. However, biodiversity remains critical for individuals, societies, and humanity as a whole.

- The Leaders' Pledge for Nature, launched in 2020, is a potential improvement on past commitments, acknowledging the need for integration, private sector finance, and shared responsibility. However, its effectiveness is undermined by the absence of several major countries.

In the next chapter

What would happen if, hypothetically, we resolved all the key biodiversity loss issues?

Chapter Seven

Exercises in thought

Regular calls to reverse environmental degradation do little to slow biodiversity loss. Follow this thought exercise to learn what would happen if the calls were heeded.

Wallowing in metrics

In May 2019, the United Nations Intergovernmental Science-Policy Platform on Biodiversity and Ecosystem Services (UN-IPBES) released a global assessment summary on biodiversity loss for policymakers. Ho hum, ho hum... excuse us while I yawn through another 40 pages of impenetrable waffle from people who, instead of saving the planet, fly around to conferences all the time, blaming the delegate with the biggest carbon footprint.

As you can see, I am way too cynical these days. Comes from having heard this all so many times before, not least during a decade working in Africa in the 1980s and 90s, where the reckoning was that only 10c of every US$ spent by the UN went onto the ground—but I digress.

As for the yawn, there is a bit of pot and black kettles here. I have been known to prattle on a bit, especially when I get onto a passionate topic. However, I don't need to say much about biodiversity loss. Sane people all around the world already know there is a serious issue of what to do about the obvious and insidious exploitation of the planet's natural resources and the degradation and extinctions that go with the disturbances. Huge numbers of voracious people on a finite planet are increasing at 8,000 souls per hour—first graders can do the maths—which means that staying within planetary limits is a wicked problem.

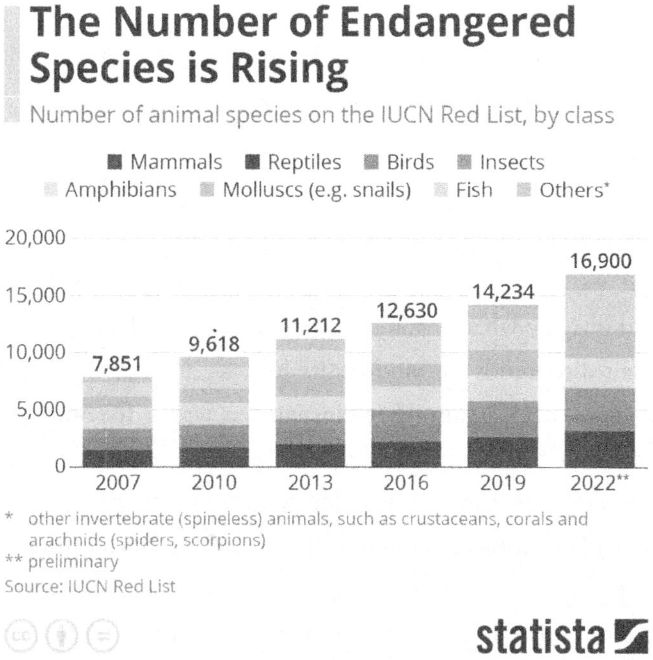

Annual growth here is roughly 16%, which if it continues would mean roughly 58,000 endangered species on the list by 2030 and 1.2 million by 2050.

Most sane people also know we repeatedly fail to find political solutions to this resource use conundrum. We keep doing the same things—for example, releasing international policy platform statements—and expect a different result. Those old enough can recall that there was quite a brouhaha over biodiversity and ecosystem services at the Rio Earth Summit in 1992. This was a milestone for the United Nations, not least because world leaders came together to discuss the environment for the first time. Periodically since then, dire warnings and policy suggestions have emerged, but as the latest assessment summary from the UN-IPBES suggests, both the condition and the prognosis are much worse today.

So rather than wallow in a bunch of metrics heading south, let's focus on biodiversity loss and try a little thought exercise around the latest summary report.

A thought exercise on biodiversity loss

Try and imagine what it would be like if, instead of listing the metrics and their dire trends for biodiversity loss, we actually did something about them. Let's assume that Einstein was wrong for once, and the same things actually delivered a result. We did some things that turned all the key metrics around. Miraculously, they are all heading in the right direction. The problem of biodiversity loss is fixed. No more loss.

To help us put this counterfactual into context, we'll do the thought exercise for Australia with some help from Guardian journalist Calla Wahlquist, who extracted from the IPBES summary the ten key points of the latest UN environment warning and what Australia must do to turn it all around. In our thought exercise, we are going to imagine that all the issues on this list are fixed because a miracle happened and there is no more biodiversity loss, indeed the diversity is returning to all the places where it used to be. The problems are sorted, and any measures are no longer heading south. Phew, this is a huge relief.

What would be different from the current and future for Australia and the rest of the world if these 10 things were resolved successfully? Here are the ten things Calla Wahlquist gleaned from the IPBES report that will happen if biodiversity loss is fixed because Samatha, the delightful star of *Bewitched*, twitched her nose, and all is well:

> Human life will be severely impacted if we do not protect biodiversity

Human life would no longer be severely impacted... from the things that biodiversity provides: clean air, clean water, good food, reliable pollination, natural pest control etc. This is all good because the key ecosystem services are back up and running. No more need to panic, except that there are a bunch of equally pressing matters that reversing biodiversity loss will not fix, in no particular order: cancer, heart disease, obesity, refugees, war, terrorism, etc. Outcome—complex but mostly good

> Species are dying at the cost of food security

Species and food security are linked because modern agriculture uses only a fraction of the species we once used for food. Variety is a universal mitigator of risk, so when the food supply comes from just a handful of species and, say, wheat blight strikes hard, then everyone is in strife. Only a fool would separate diversity from our means of production, and an even bigger fool would toy with 8 billion hungry souls for the sake of a koala. Outcome—good but not necessarily for the reasons stated.

We should support Indigenous knowledge and land management

Indigenous knowledge is remarkable, given its history, which has been passed down through generations for so long. The land management options that first people know and would prefer to use are exceptional. Now, what we are about to say is not politically correct, and we do know that indigenous people have much to say about the way the environment works, **but**, bless them, what they know came from a world with very few people. What they know came from a world without cars, roads, planes, and a food supply designed around technology. Outcome—useful with helpful learnings but from another time.

Native people's land rights

Land rights do matter because they come with a social licence to use land and the resources and production capacity it holds. But as with indigenous knowledge, returning land rights is good but does not solve the problem. The indigenous systems didn't have close to 8 billion mouths to feed. Outcome—morally correct but will not solve the problem of biodiversity loss.

Reforestation for carbon sequestration could harm biodiversity

Harming biodiversity from carbon sequestration in trees is a weird one. If we choose not to grow trees, then that's one source of carbon sequestration that we have to forgo. Given the emissions levels, we will need every single option to clean up the climate mess—if that is even possible. So we solve one problem, but we leave ourselves vulnerable to another, climate change, that is equally pressing. Plus, any trees are likely to enhance biodiversity over no trees even if they are not the tree species that were cut down in the first place. Outcome—not even logical.

We have to protect green spaces in cities

It is critical to maintain **green spaces in cities** for general public health. That green space also provides for biodiversity is a bonus, not a driver. Cities are not where biodiversity lives, it's where the people live. Green spaces are a good thing to do but are hardly material to the bigger problem of how to generate 22 trillion kilocalories a day for hundreds of years. Outcome—good but not material to the real issues of biodiversity loss.

We are losing a lot of water to agriculture

When there are 8 billion people, there is a huge demand for food. In most places, successful production will require the addition of **water for agriculture** either through the growing cycle or as food is processed. We are not losing water to agriculture**; we have to have water for agriculture, period.** If we reduce water use in food production, the food security risk gets more acute, not less. Outcome—makes the real problems more acute.

We need to ditch subsidies that encourage environmentally harmful practices.

Ditching subsidies is a no-brainer. Whatever humans have done over the generations, most decisions are based on who makes a dollar. All the great wars, through to all the tiny arguments over who gets the family inheritance, are ultimately based on money, and humanity has allowed interest groups to determine where some of that money goes to the detriment of the environment. So, this one might make a difference to rates of biodiversity loss. Outcome—good.

We will lose all the coral if we can't keep climate change below 2C

In the counterfactual scenario where the temperature rise stays below 2 degrees Celsius, some of the **coral reefs survive,** and a proportion of biodiversity is saved. This means a few tourists can visit iconic places and enjoy themselves, and the rest of us can sleep easy knowing that Nemo is both found and saved. But in the grand scheme, this is just a tiny piece of a gigantic puzzle. Outcome—good for part of the economy and a small fraction of the biodiversity.

We need to take an ecosystem-based approach to manage fisheries

Fish are a critical part of the diet of half the planet. It has always been thus. The majority of people have lived close to the water for obvious reasons. **Managing fisheries as ecosystems** will help retain fish as a critical protein source, but again it's only a single piece of a much larger puzzle. It is a bigger piece than coral reefs and relevant because it is provisioning ecosystem service and is not primarily aesthetic but still only a piece. Outcome—good for part of the economy.

Species richness in wildflower meadows can range from a few dozen to over 100 species per square meter in highly diverse areas

A better list of things to do.

As you can see, a mindful sceptic would have some doubts and might conclude that Calla Wahlquist's list extracted from the IPBES summary is uninspiring at best. Fixing biodiversity loss in these ways barely touches the surface of the bigger problem of supporting 8 billion people without further depleting natural resources.

Our curiosity and critical thinking led us to a shorter list of critical environmental issues that are **not fixed** after the 10 actions in the counterfactual thought process came to pass thanks to Samantha's twitching nose and the miracle of biodiversity loss coming to an abrupt end. Here are some of the salient items.

- The fact that the FAO determined that 40% of global soils are degraded
- We have a critical shortage of accessible nitrogen to fertilise crops, and what we have is made with fossil fuels
- Soil erosion impacts a significant proportion of production landscapes
- Many agricultural systems will not cope with the already changing climate

And we could go on, given we are already in ecological overshoot. In other words, the world still has a swag of mission-critical environmental issues that are not even touched by fixing up biodiversity loss. And this is a key problem for those who want to conserve biodiversity, including the UN-IPBES. Their critical message is that **biodiversity loss needs to stop.** True enough, for a whole range of reasons, not least the ecosystem services and soil health that biodiversity delivers. But halting biodiversity loss is just one of a swag of important messages that people will need to heed to get through the global population spike and still have a habitable planet.

There is no doubt that the loss of biodiversity is critical and not, as so many of these report-style documents assert, because of what it is but **because of what it does.** Biodiversity provides essential services, and diversity confers resilience in how these services perform. Keeping and enhancing biodiversity reduces risk, and humanity will depend on managing ever-increasing food security risks.

The problem here is that the core of the UN-IPBES argument is still that we must conserve, that we must stop, and that we must retain things as they were. This is not possible anymore. The world has changed, and we must change with it. We have to reset the way we think about biodiversity loss to understand the real reasons for loss, put much more effort into soil biodiversity, and become more united to resolve the issues.

A few UN reports are not enough. Let's have a look at another popular solution to slow or stop biodiversity loss: rewilding.

Is rewilding the solution to save the planet?

I watched an episode of Australian Gogglebox—where people watch TV watching people watching TV—bizarre but hilarious. The Australian viewing public at home on their couches is so funny; their wisecracks and personalities light up the screen far more than the shows they are watching. One of the programs was David Attenborough's *A Life on Our Planet.* All the Goggleboxers, despite different backgrounds and cultures, had fear in their eyes. Their concern over the messages was palpable. They all recognized what ecologists have known for a long time:

that the disturbance from agriculture, fisheries and human land use have decimated wildlife populations across the planet.

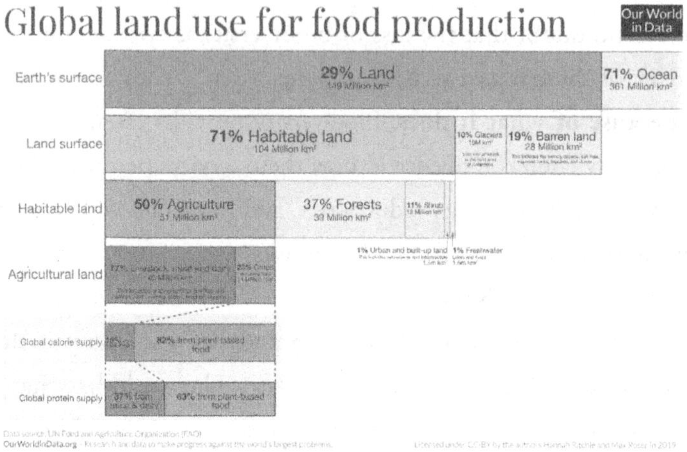

Half the habitable land is used for agriculture, the majority for grazing animals.

Adding climate change to land-use change means that even if we haven't physically altered habitats, the climate will do that for us. Recall that the geological record suggests that four of the five previous mass extinctions of biodiversity on the planet resulted from climate change. Over millennia in those instances, but climate nonetheless. The message that Sir David delivered, and he must be commended for finally coming off the fence and telling his audience the truth after so many years of wowing them with the glory of nature, is that **life on earth is in big trouble**.

Natural capital, the engine of all our food production and the iconic biodiversity that drives it is being depleted everywhere. And in some places, this capital is mined by the production systems and land uses humans have developed. And as we have seen, biodiversity is declining at rates equivalent to a mass extinction event. This biodiversity loss is especially important in soils, given that biology drives the transfer of nutrients to plants. There is also the matter of 22 trillion kilocalories per day, the energy needed in the food supply to keep 8 billion people from going hungry.

Sir David's solution to these problems is **to rewild a third of the planet**. This an excellent catchphrase that is attractive and accessible to many people. It implies that some of the former wilderness that humanity has flattened be returned to something close to its original state. So, is rewilding the solution? First, we need to know what Sir David meant by rewilding.

What is rewilding?

Rewilding is pretty much what it says on the tin. That is a return to the wild by removing human influences and letting nature take care of itself, nudged along by the reintroduction of megafauna, such as large grazers or large carnivores, to restore ecosystem processes. More formally, Pettorelli et al. (2018) define rewilding as

> *"the reorganisation of biota and ecosystem processes to set an identified social–ecological system on a preferred trajectory, leading to the self- sustaining provision of ecosystem services with minimal ongoing management."*
>
> Pettorelli, N., Barlow, J., Stephens, P. A., Durant, S. M., Connor, B., Schulte to Bühne, H., ... & du Toit, J. T. (2018). Making rewilding fit for policy. Journal of Applied Ecology, 55(3), 1114-1125.

Rewilding intervention might try to mirror the ecology and species compliments that were present in the past, but reorganisation proceeds according to natural processes to an uncertain endpoint or even to no endpoint at all. So long as the ecosystem is resiliently providing ecosystem services and this is ongoing, rewilding has succeeded. In this sense, rewilding differs from the related idea of ecological restoration, which requires a prescribed endpoint by being open-ended.

Eucalyptus woodland in eastern Australia is already wild and shaped by fire events but lacks the megaherbivores

Rewilding, '*the practice of returning areas of land to a wild state, including the reintroduction of animal species that are no longer naturally found there*' is a courageous idea. And to suggest this happens over a third of the global land area is a huge ask. As Sir David says, rewilding begins by halting the processes that are degrading the system. It will mean stopping or modifying the activities that degrade natural capital and then putting extraordinary effort into rebuilding natural habitats.

Why do this?

A third of the globe returned to the wild should slow rates of biodiversity loss. This will not guarantee the survival of any given species; the iconic rare and endangered will still need individual attention, but enhanced preservation of the many will maintain ecosystem services with follow-on effects for climate mitigation and water balance.

Feeding ourselves on a rewilded planet

Human instinct is to survive, and if the food begins to run out, we will do whatever it takes. This has been the case throughout human history. Most of our conflicts are about resources of one sort or another, even if they are disguised as political agendas or megalomaniac tendencies. Throwing resources into rewilding through rebuilding nature is only possible if humans are also in good shape.

Sir David tempers his argument for rewilding with the requirement to grow food. In the documentary, drones collect nuts from rainforest trees, but that will not be enough. There is a 30-year window to grow and then maintain an extraordinary level of food supply distributed efficiently across the world before the demographic transition passes and there are fewer people to support.

Additionally, wild areas generate provisioning services, such as drones harvesting nuts from the forest; regulating services, such as controlling climate and disease; supporting services, such as nutrient cycles and oxygen production; and cultural services, such as spiritual and recreational benefits. When all these services are integrated into the landscape, they benefit production, too.

Rewilding as part of the solution

The scope of rewilding, as proposed by Sir David, is ambitious. It is a global idea—something like the 'man on the moon' audacity of the early 1960's. It could slow biodiversity loss, but there is a caveat. Rewilding has to include a rewilding of the production system, too. Not just the spare land that we don't currently use. It has to be an integrated approach to food production that maintains ecological processes across the landscape. Rewilding areas not under production will do little to alter biodiversity loss rates. We already know that the reserve systems don't do enough to halt or even slow biodiversity loss, part of Sir David's original lament. So, wildlife conservation and nature management have to be integrated into the production system, making them an integrated connected unit across landscapes, ideally the entire planet.

Is rewilding a solution to biodiversity loss? There is no doubt that loss will continue at mass extinction rates unless something serious is done. Reserves are not cutting it, nor will any species-level initiatives. Wild animals and plants will need much more space to survive than we have left them.

Rewilding will be the answer in some areas, especially those that have been degraded by human agricultural activity but were not suitable for food production in the long term or those areas that need a rest from production. Abandoned agricultural land is a great place to start. It may feel unattainable, especially in the form Sir David Attenborough suggests as a rewilding of one-third of the earth's land area. But the reality is that humanity has met equally unimaginable challenges in the past, and this one we have to achieve for any sustainable future.

The big question is, if a third of the planet is rewilded, even with multiple use options, can the whole system deliver nutritious food equivalent to 22 trillion kilocalories daily for 100 years?

Rewilding is nice, a sop, and has to happen, but it is a mitigation, not a solution. It is a bit like taking paracetamol for a headache.

Key Points

- Despite repeated warnings from the UN and other organizations since the 1992 Rio Earth Summit, biodiversity loss continues to worsen, leading to scepticism about the effectiveness of global assessment reports and policy recommendations.

- A thought exercise exploring the hypothetical resolution of key biodiversity loss issues reveals that while some outcomes would be positive, many critical environmental problems, such as soil degradation, nitrogen shortage, and the vulnerability of agricultural systems to climate change, would remain unaddressed.

- Halting biodiversity loss is crucial, but it is only one of many issues that must be tackled to ensure a habitable planet for the growing human population. Given the extent of global change, the focus on conservation and maintaining the status quo is no longer feasible.

- Rewilding, which involves returning land areas to a wild state and reintroducing animal species, can be part of the solution to slow biodiversity loss. However, it must be integrated with a rewilding of the food production system to maintain ecological processes across the landscape and ensure sufficient food production for the global population while preserving biodiversity.

In the next chapter

It might be good for a sound bite to conflate the two, but biodiversity and conservation differ. They are not the same thing.

Chapter Eight

Why rare species conservation might not matter

Unfortunately, there are more pressing issues than rare species conservation for feeding everyone well.

What humans need

Many people see the solution to biodiversity loss as identifying and saving, through various direct efforts, specific species, especially those that are already rare and at risk of extinction. In this paradigm, we happily conflate conservation and biodiversity to mean the same thing. This is a huge mistake.

The needs wants, and desires of 8 billion people are at odds with rare species conservation. Humanity has already pushed and breached critical planetary boundaries, and we live precariously to feed ourselves, let alone find space for nature in the planet's habitable regions. Our energy use and resource consumption produce waste that also impacts the uninhabitable parts of the earth, from the deep oceans to the Arctic tundra. Humans genuinely are a phenomenon.

So, does rare species conservation have a chance? A slim one, given that 96% of the mammalian biomass on earth is humans and their livestock, and our pledges are failing to reverse biodiversity loss.

Biodiversity and conservation are not the same

Surprising as it may seem, biodiversity and conservation are not the same. Most importantly, biodiversity does not mean biological conservation. Biodiversity does not mean rare or endangered either, and it is not a synonym for an environmental value of your choice.

As the original International Biodiversity Convention agreed way back in 1992, the formal definition of biodiversity is

> *the variability among living organisms from all sources including, inter alia, terrestrial, marine and other aquatic ecosystems and the ecological complexes of which they are part; this includes diversity within species, between species and of ecosystems.*

Rather than be so cumbersome, most biologists define biodiversity as the

> *totality of genes, species and ecosystems of a region*

or more simply still, from the title of Ed Wilson's 1992 book...

In the environmental context, conservation is protecting, preserving, managing, and restoring natural environments and their ecological communities. The goal is to maintain biodiversity, ecosystem functions, and the sustainable use of natural resources for current and future generations. In other words, it is the practical application of a value. Conservation is a dynamic field that continually evolves in response to new scientific understanding, changing environmental conditions, and societal needs. It aims to balance the preservation of nature with sustainable human development, recognizing the interconnectedness of ecological and human systems. It is not biodiversity, even when biodiversity is what is being conserved.

Though these definitions are very clear—biodiversity is about variety and conservation about values—biodiversity has morphed in the eyes of the public and the media to mean conservation of rare things or just conservation, period. This shameless hijacking of one technical term to mean something else to promote a specific agenda is all too common. I have lamented this general topic on my blog many times, and I even went on about it in previous books, *Awkward News for Greenies* (2009), *Missing Something* (2012) and the upcoming *More than the Rare Things*.

Biodiversity, the *diversity of life*, is the consequence of evolution and ecology—the outcomes of interactions between organisms in time and place. One outcome is an extraordinary profusion of organisms. There are so many different types that there is no single catalogue. Indeed, Australian taxonomists, the scientists who describe and catalogue the variety of life, estimate that $824m is needed over the next 25 years to complete the biodiversity record for the island continent. Meanwhile, the New South Wales government has set a target of zero extinctions of native wildlife in the state's national parks estate, the first time an Australian government has set the goal. Zero extinction is impossible to prove, not for want of on-ground effort or technical skill, but because the species catalogue is incomplete. But, I digress.

Species are just one of the outcomes of evolution contained within '*the diversity of life*'. Biodiversity is about biology, what it is and what it does, and not just the number of different organism types. In contrast, conservation is a value, not an object, thing, or process. And the key to understanding this particular human perception is that, in conservation, **rare is more valuable than common**.

I was lucky to take this photo of a young lioness in Hwange National Park, Zimbabwe, in the early 1990s. She is no longer alive as wild lions live for 16 years at best.

However, humans also value clean air, clean water, and food. These essentials are delivered primarily by the actions of biodiversity, particularly the common species. Indeed, there have been very few times when rare species have been pivotal to these essential services. A little thought, and this truth is clear. A rare organism is infrequent by definition, and it cannot easily influence services by biomass or weight of numbers. The only way a rare species can affect services is to engineer the environment, such as a beaver building a dam that alters the transfer of nutrients and water across the landscape. Rare, as in endangered or threatened, rarely matters for ecosystem services.

Why rare species conservation might not matter

The reason for distinguishing between rare and common species in the context of human needs from nature is **ecology**.

> *Ecology (from Greek: οἶκος, "house" and -λογία, "study of") is the study of the relationships between living organisms, including humans, ...and their physical environment*

The interactions between organisms capture and convert energy from the sun into biomass (plants) and then transfer that energy through systems (plants and animals) are moderated by the availability of water and sunlight. This ecology is what keeps people alive. Ecology gives humans food, clean water, and many valuable products—collectively, we call this natural capital, natural resources, or ecosystem services.

This truth is fundamental to human existence. We only exist through ecology, for even the most intensive farming systems rely on interactions between microbes and plants.

A fixation on iconic species with fur or feathers is all very well, and there is moral, cultural and existential value in rare species conservation.

Trying hard to save iconic species is a focus that might help us avoid insanity and help bring us back from the brink. In this regard, all conservation is worth it.

The problem is that rare species conservation is not the highest priority. The species combinations that deliver and maintain critical services in the landscape need our full attention. The common species allow humans to persist, especially those in agricultural landscapes. And these species are not often those in danger of extinction. But their abundance and distribution are changing. As we saw in Chapter 3, insects in a selection of nature reserves in Germany have declined by 75% in a generation. This level of decline is dramatic. If it happened to global stock exchanges, people would panic. There would be a run on the banks, a great depression would be declared, and the fragile financial house of cards would collapse along with the six-continent food supply chain.

I took this photo in the late 1980s of a White Rhino mother and mature calf in Hwange NP, Zimbabwe. All the big mammals of Africa will join the rhinos in the rare species conservation category should the projected human population growth on the continent continue.

Given what insects do to transfer nutrients, ensure food chain stability, and pollinate—just three of the critical ecosystem services—we should also be panicked by these changes. The erosion of ecosystem services is more critical to humanity than the loss of rare species.

A shift from the rare to the common.

Promotion of the common over the rare makes me sound like a rabid developer, but remember, there are 8 billion people who need food, water, and shelter. There is an unprecedented demand for ecosystem services when many are dramatically declining. If all the critically endangered species on the IUCN *Red List of Threatened Species* were to go extinct, it would be a cultural and moral disaster. Such a tragedy would leave a lasting sadness. But it would not be the end of the world.

I know this sounds callous, but humanity would still till the soil, plant grains, rear livestock and eat without the rare species. Should the rarities be lost, agriculture would not because nature has an uncanny knack for persistence. There would still be ecology because nature will always occupy available space and continue to have organisms interacting to transfer energy and nutrients around.

The persistence of nature is critical and brings us to the pivotal diversity of life, the biodiversity that humanity relies on without even realising it—**the biodiversity that we must have is in the soil.** The tens of thousands of species that live in each square metre of soil are essential for the nutrient transfer, water retention and soil health that delivers crops and livestock production. Humanity might live in high-rise blocks, drive electric cars and send rockets into space, but all this depends on food grown in soil.

Just a few know this. Most people have no reason to think about this version of biodiversity. After all, food comes from the fridge or the supermarket, depending on which family member you are. Perhaps the attention to rare species conservation is the subconscious feeling of our intimate connection and reliance on nature and expressing that connection through something our conscious brains can grasp.

I took this photo in the late 1980s of a dung beetle burying elephant dung in Hwange NP, Zimbabwe.

Rare species conservation does not matter for feeding everyone well. There are more pressing issues that we have to fix or risk even greater catastrophe than the loss of some iconic species. Humanity must increase food production by 2% for 30 years to feed everyone well. Then, these increases must be maintained long enough for the demographic transition to pass. If we fail to produce enough food, it will destabilise the economic and social systems before we can replace them, and we will crash and burn.

Hope for avoiding catastrophe is to focus on soils. We also need to focus on the natural capital of our environment to ensure that, along with other technical fixes, we can grow enough food so that at least we don't have a war over our primary resource. This sounds dramatic, but there is no obvious argument against it. Any extinction is tragic, especially those that are preventable. However, focusing on rare species conservation at the expense of critical diversity that delivers services, especially in soil, is short-sighted.

Saving the koala is great, but not if food is insufficient to feed all the people and their pets. Halting biodiversity loss for the many, especially in the soil, makes much more sense.

Think about it.

Key Points

- Biodiversity and conservation are not the same; biodiversity refers to the variety of life, while conservation is a human value. Rare species, due to their low numbers or biomass, cannot significantly influence ecosystem services, whereas common species play a crucial role in providing essential services.

- While rare species conservation has moral, cultural, and existential value, it should not be prioritized over the species combinations that deliver and maintain critical ecosystem services, particularly in agricultural landscapes. The erosion of ecosystem services is more critical to humanity than the loss of rare species.

- Soil biodiversity is essential for nutrient transfer, water retention, and soil health, enabling crop and livestock production. Focusing on soils and the natural capital of our environment is crucial to ensure that we can grow enough food to prevent resource-driven conflicts

- The primary focus should be on increasing food production

by 2% annually for the next 30 years and maintaining these increases long enough for the demographic transition to pass. Prioritizing rare species conservation at the expense of critical diversity that delivers ecosystem services, especially in soil, is short-sighted in the face of the pressing need to feed the growing human population.

Chapter Nine

Postscript

There are thousands of books on positivity. Every week, new ones emerge that squeeze every last nuance out of 'get up and go'. Amazon lists over 80,000 just on positivity, so the concept must sell.

One of the common themes is to think big. Upsize your ambition and your fries; truly exceptional things can happen. Your thinking, the ideas you can conjure and your cholesterol levels can all expand if you upsize enough. It is tempting.

In the early 2000s, I had great plans for a venture of mine, Biotrack Australia, that monitored environmental performance for a fee. The technique was to sample biodiversity as a signature of habitat and, with judicious comparison, a measure of the status of natural capital. We would sample invertebrates and use their diversity to detect the impact of insecticides, the management of arable land, the effects of forestry on ecosystem services and the overall health of streams. It was a cool tool with, for the time, neat applications of image and database technologies to speed up the specimen identification process.

Rather than leave this capability in the ivory tower, I incorporated the idea and ran it as a commercial operation. There were visions of expansion from a few local government, research, and mining company contracts to global dominance. Fuelled by much hard work and some great staff, the technology and the business improved. We even set up a branch in Botswana (long story). I imagined a neon sign atop a high-rise in North Sydney beaming out the company logo across the iconic harbour.

Then, the cash ran out. Instead of expanding to the world, Biotrack downsized. The fees did not match the level of interest potential clients had in finding out just how well their environment was performing. And you can see why. Only if a company has to report its externalities it does not need to measure them. Knowledge has to make a material difference to the bottom line, and even then, decision-makers need to feel the immediate benefit of investing in it.

So Biotrack spent what we had left on an all-out search for clients who would profit from knowing their environmental performance and failed to find any. Soon after the technology and IP were sold for a modest amount. There was no neon this time.

Here is the younger model spruiking the benefits of technology to use biodiversity as a monitoring tool.

The post-mortem was brief. We decided that we'd experienced an all-too-familiar problem for commercial start-ups: some success prompted growth in advance of the cash flow to support it. In short, we were undercapitalised. Our resources did not match the grand vision.

This was a satisfactory explanation at the time and supported the positive grand vision. Given that nobody has ever won the lottery without a ticket, we felt we had given it a fair go but didn't have the necessary luck.

Perhaps I should have done better. Rather than using biodiversity as an indicator of externalities, I should have measured it for its own sake. With the ability to sample and process tens of thousands of specimens in a few weeks, which by now would be days or even hours, Biotrack could have provided all the information ever needed on biodiversity loss and any gains that might accrue from actions to save it.

But here is the thing. Biodiversity does not make money, at least not directly. Sure it underpins a raft of ecosystem services that support the entire human endeavour, but that comes from natural capital, the resource humans were put on the planet to exploit. It was trendy to try to commercialise research, only that only happens if there is a market.

I sometimes wonder if, instead of giving in to my entrepreneurial gene, I had stayed as an academic hiding behind the technical details and the notion that biodiversity was just ecology and evolution. Unfortunately, the ivory tower metaphor is real. Universities are bastions of highbrow thinking disconnected from real-world problems and everyday concerns. So, whilst I may have added a little to pure knowledge, with its theoretical and abstract thinking, it would have done little to meet the challenge of biodiversity loss.

Little has changed in the last 20 years. Conservationists are focused on rare endangered species, academics still need clarification on their theories and, in the real world, exploiting natural resources is an economic necessity.

Biodiversity loss remains an inevitability.

One of the images captured by the Biotrack team to assist with rapid identification of insect species; in the case the ant, Solenopsis invicta

Some of the research output from Biotrack

- Dangerfield, J. M., & Pik, A. J. (1999). The educational value of an all taxa biodiversity inventory. *Journal of Biological Education, 33*(2), 76-83.

- Oliver, I., Pik, A., Britton, D., Dangerfield, J. M., Colwell, R. K., & Beattie, A. J. (2000). Virtual biodiversity assessment systems. *Bioscience, 50*(5), 441-450.

- Beattie, A. J., Dangerfield, J. M., Pik, A. J., & Oliver, I. (2002). Understanding more of biodiversity: The role of information technology. *Today's Life Science, 14*(3), 30-34.

- Pik, A. J., Dangerfield, J. M., Bramble, R. A., Angus, C., & Nipperess, D. A. (2002). The use of invertebrates to detect small-scale habitat heterogeneity and its application to restoration practices. *Environmental Monitoring and Assessment, 75*, 179-199.

- Pik, A., Beattie, A., & Dangerfield, M. (2002). BioTrack R simplifies monitoring of invertebrates on restored sites (Australia). *Ecological Restoration, 20*(1), 67-68.

- Dangerfield, J. M., Pik, A. J., Britton, D., Holmes, A., Gillings, M., Oliver, I. A. N., ... & Beattie, A. J. (2003). Patterns of invertebrate biodiversity across a natural edge. *Austral Ecology, 28*(3), 227-236.

- Oliver, I., Holmes, A., Dangerfield, J. M., Gillings, M., Pik, A. J., Britton, D. R., ... & Beattie, A. J. (2004). Land systems as surrogates for biodiversity in conservation planning. *Ecological Applications, 14*(2), 485-503

Science sources

As with all the Mindful Sceptic Guides, the literature on food security is vast. We recommend using Google Scholar with the keyword 'food security' to find material that tickles your curiosity.

Rather than a comprehensive list of sources, check out our Evidence Reviews, which you can access by clicking on or scanning the image below.

Mindful Sceptics

evidence reviews

scan or click the QR code for a deeper dive into the science sources for the Mindful Sceptic Guides

SCAN ME

About the Authors

Mark is an ecologist, not a green one. He has built, run and sold companies, held academic posts at four universities, advised governments on environmental matters, has a PhD and still plays too much golf. His ecological research ended up in more than 70 peer-reviewed research papers and over 80 technical reports, and he supplied peer-review for more than a dozen international science journals. He lived in Africa for a decade, where he researched termites and millipedes, and once walked into a lion who snarled at him. Much of his career has been in Australia, where he won an Outstanding Teacher award from Macquarie University.

His consultancy practice advises commercial and public sector clients on assessing, monitoring, and evaluating environmental impacts, especially in carbon asset and liability management. His non-fiction books are about awareness of humans' role in nature and, as Paul Sorol, he chronicles and writes fiction for a change.

Link in bio

Chris is an ecologist, a restoration one. A life-long mindful sceptic, Chris has happily landed as an ecological practitioner in his bush regeneration business. When not passionately managing land, being awarded Landcarer of the Year, trawling the evidence on nutrition, diet and health or carefully advising NGOs and government, he grows plants in his commercial nursery.

MINDFUL SCEPTIC GUIDES

Abatement

Biodiversity loss

Escaping the Malthusian trap

Farmers

Food security

Food, people and soil

Growing enough food

Healthy and sustainable diet

Sustainability

The soil carbon debt

mindfulsceptics.info

Made in the USA
Monee, IL
04 November 2026

67380066R10075